Interactive Student Edition

Reveal MATH®

Course 3 • Volume 2

Mc
Graw
Hill

Cover: (l to r, t to b) David G Hemmings/Moment/Getty Images, sweetmoments/ E+/Getty Images, DeadDuck/E+/Getty Images, oxign/E+/Getty Images

mheducation.com/prek-12

Copyright © 2020 McGraw-Hill Education

Send all inquiries to:
McGraw-Hill Education
STEM Learning Solutions Center
8787 Orion Place
Columbus, OH 43240

ISBN: 978-0-07-899718-1
MHID: 0-07-899718-6

Reveal Math, Course 3
Interactive Student Edition, Volume 2

Printed in the United States of America.

12 13 14 15 16 SWI 25 24 23 22

Contents in Brief

Module 1 Exponents and Scientific Notation

2 Real Numbers

3 Solve Equations with Variables on Each Side

4 Linear Relationships and Slope

5 Functions

6 Systems of Linear Equations

7 Triangles and the Pythagorean Theorem

8 Transformations

9 Congruence and Similarity

10 Volume

11 Scatter Plots and Two-Way Tables

Reveal Math® Makes Math Meaningful...

Interactive Student Edition

Learning on the Go!

The flexible approach of *Reveal Math* can work for you using digital only or digital and your *Interactive Student Edition* together.

Student Digital Center

...to Reveal YOUR Full Potential!

Reveal Math® Brings Math to Life in Every Lesson

Reveal Math is a blended print and digital program that supports access on the go. You'll find the *Interactive Student Edition* aligns to the Student Digital Center, so you can record your digital observations in class and reference your notes later, or access just the digital center, or a combination of both! The Student Digital Center provides access to the interactive lessons, interactive content, animations, videos, and technology-enhanced practice questions.

Write down your username and password here

Username: _____

Password: _____

Go Online!
my.mheducation.com

Web Sketchpad® Powered by The Geometer's Sketchpad®- Dynamic, exploratory, visual activities embedded at point of use within the lesson.

Animations and Videos – Learn by seeing mathematics in action.

Interactive Tools – Get involved in the content by dragging and dropping, selecting, and completing tables.

Personal Tutors – See and hear a teacher explain how to solve problems.

eTools – Math tools are available to help you solve problems and develop concepts.

Module 1
Exponents and Scientific Notation

e Essential Question
Why are exponents useful when working with very large or very small numbers?

What Will You Learn? 1

Lesson   **1-1**   **Powers and Exponents** 3
Explore   Exponents

1-2   **Multiply and Divide Monomials** 13
Explore   Product of Powers
Explore   Quotient of Powers

1-3   **Powers of Monomials** 25
Explore   Power of a Power

1-4   **Zero and Negative Exponents** 33
Explore   Exponents of Zero
Explore   Negative Exponents

1-5   **Scientific Notation** 43
Explore   Scientific Notation

1-6   **Compute with Scientific Notation** 55

Module 1 Review 63

Module 2
Real Numbers

Essential Question
Why do we classify numbers?

		What Will You Learn?	67
Lesson	2-1	**Terminating and Repeating Decimals**	**69**
		Explore Terminating Decimals	
	2-2	**Roots**	**79**
		Explore Find Square Roots Using a Square Model	
	2-3	**Real Numbers**	**91**
		Explore Real Numbers	
	2-4	**Estimate Irrational Numbers**	**101**
		Explore Roots of Non-Perfect Squares	
	2-5	**Compare and Order Real Numbers**	**111**
		Module 2 Review	123

Module 3
Solve Equations with Variables on Each Side

e Essential Question

How can equations with variables on each side be used to represent everyday situations?

What Will You Learn? 127

Lesson   **3-1** **Solve Equations with Variables on Each Side** 129
Explore Equations with Variables on Each Side

   3-2 **Write and Solve Equations with Variables on Each Side** 137
Explore Write and Solve Equations with Variables on Each Side

   3-3 **Solve Multi-Step Equations** 145

   3-4 **Write and Solve Multi-Step Equations** 151
Explore Translate Problems into Equations

   3-5 **Determine the Number of Solutions** 159
Explore Number of Solutions

   Module 3 Review 169

Module 4
Linear Relationships and Slope

e Essential Question
How are linear relationships related to proportional relationships?

		What Will You Learn?	173
Lesson	**4-1**	**Proportional Relationships and Slope**	175
		Explore Rate of Change	
	4-2	**Slope of a Line**	191
		Explore Develop Concepts of Slope	
		Explore Slope of Horizontal and Vertical Lines	
	4-3	**Similar Triangles and Slope**	205
		Explore Right Triangles and Slope	
	4-4	**Direct Variation**	213
		Explore Derive the Equation $y = mx$	
	4-5	**Slope-Intercept Form**	225
		Explore Derive the Equation $y = mx + b$	
	4-6	**Graph Linear Equations**	237
		Module 4 Review	247

Module 5
Functions

e **Essential Question**

What does it mean for a relationship to be a function?

What Will You Learn? 251

Lesson   **5-1**   **Identify Functions** 253
Explore   Relations and Functions

5-2   **Function Tables** 263
Explore   An Introduction to Function Rules

5-3   **Construct Linear Functions** 273

5-4   **Compare Functions** 285
Explore   Comparing Properties of Functions

5-5   **Nonlinear Functions** 293
Explore   Linear and Nonlinear Functions

5-6   **Qualitative Graphs** 305
Explore   Interpret Qualitative Graphs

Module 5 Review 313

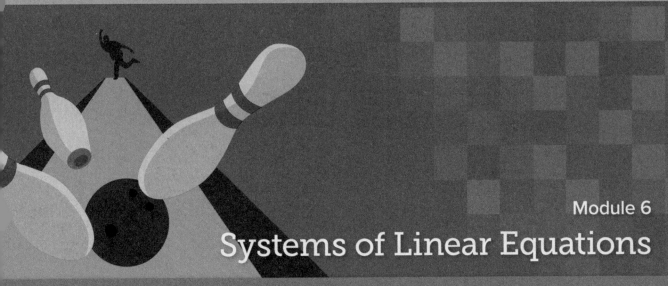

Module 6

Systems of Linear Equations

e Essential Question

How can systems of equations be helpful in solving everyday problems?

What Will You Learn? 317

Lesson   **6-1**   **Solve Systems of Equations by Graphing**................................ 319

Explore   Systems of Equations

6-2   **Determine Number of Solutions** 331

Explore   Systems of Equations: Slopes and *y*-Intercepts

6-3   **Solve Systems of Equations by Substitution**............................ 341

Explore   Solve Systems of Equations by Substitution

6-4   **Solve Systems of Equations by Elimination**................................ 351

Explore   Solve Systems of Equations by Elimination

6-5   **Write and Solve Systems of Equations**.. 363

Module 6 Review 375

TABLE OF CONTENTS

Module 7
Triangles and the Pythagorean Theorem

e Essential Question
How can angle relationships and right triangles be used to solve everyday problems?

What Will You Learn? 379

Lesson   **7-1**   **Angle Relationships and Parallel Lines** 381
Explore   Parallel Lines and Transversals

7-2   **Angle Relationships and Triangles** 393
Explore   Angles of Triangles
Explore   Exterior Angles of Triangles

7-3   **The Pythagorean Theorem** 405
Explore   Right Triangle Relationships
Explore   Proof of the Pythagorean Theorem

7-4   **Converse of the Pythagorean Theorem** 417
Explore   Prove the Converse of the Pythagorean Theorem

7-5   **Distance on the Coordinate Plane** 423
Explore   Use the Pythagorean Theorem to Find Distance

Module 7 Review 429

Module 8
Transformations

e Essential Question
What does it mean to perform a transformation on a figure?

What Will You Learn? **433**

Lesson **8-1 Translations** **435**
Explore Translate Using Coordinates

8-2 Reflections **445**
Explore Reflect Using Coordinates

8-3 Rotations **455**
Explore Rotate Using Coordinates

8-4 Dilations **465**
Explore Dilate Figures on the Coordinate Plane

Module 8 Review **475**

TABLE OF CONTENTS

Module 9

Congruence and Similarity

e Essential Question

What information is needed to determine if two figures are congruent or similar?

What Will You Learn? **479**

Lesson   9-1   **Congruence and Transformations** **481**
         Explore   Congruence and Transformations

         9-2   **Congruence and Corresponding Parts** **493**

         9-3   **Similarity and Transformations** **501**

         9-4   **Similarity and Corresponding Parts** **513**
         Explore   Similar Triangles
         Explore   Angle-Angle Similarity

         9-5   **Indirect Measurement** **523**
         Explore   Similar Triangles and Indirect Measurement

         Module 9 Review **529**

Module 10
Volume

e Essential Question
How can you measure a cylinder, cone, or sphere?

		What Will You Learn?	533
Lesson	10-1	Volume of Cylinders	535
		Explore Volume of Cylinders	
	10-2	Volume of Cones	543
		Explore Volume of Cones	
	10-3	Volume of Spheres	551
	10-4	Find Missing Dimensions	559
	10-5	Volume of Composite Solids	567
		Module 10 Review	575

Module 11
Scatter Plots and Two-Way Tables

e Essential Question
What do patterns in data mean and how are they used?

		What Will You Learn?	579
Lesson	**11-1**	**Scatter Plots**	**581**
		Explore Scatter Plots	
	11-2	**Draw Lines of Fit**	**591**
		Explore Lines of Fit	
	11-3	**Equations for Lines of Fit**	**599**
	11-4	**Two-Way Tables**	**609**
	11-5	**Associations in Two-Way Tables**	**619**
		Explore Patterns of Association in Two-Way Tables	
		Module 11 Review	**629**

Triangles and the Pythagorean Theorem

e Essential Question

How can angle relationships and right triangles be used to solve everyday problems?

What Will You Learn?

Place a checkmark (✓) in each row that corresponds with how much you already know about each topic **before** starting this module.

KEY

⬛ — I don't know.   ◗ — I've heard of it.   ⭐ — I know it!

	Before			After		
	⬛	◗	⭐	⬛	◗	⭐
classifying angle pairs						
finding missing angle measures using angle pair relationships						
finding missing angle measures using relationships between interior and exterior angles of triangles						
finding the length of a leg or the length of the hypotenuse in a right triangle using the Pythagorean Theorem						
determining whether a triangle is a right triangle using the converse of the Pythagorean Theorem						
finding distance on the coordinate plane						

📖 Foldables Cut out the Foldable and tape it to the Module Review at the end of the module. You can use the Foldable throughout the module as you learn about triangles and the Pythagorean Theorem.

What Vocabulary Will You Learn?

Check the box next to each vocabulary term that you may already know.

☐ alternate exterior angles

☐ alternate interior angles

☐ converse

☐ converse of the Pythagorean Theorem

☐ corresponding angles

☐ exterior angles

☐ hypotenuse

☐ interior angles

☐ legs

☐ line segment

☐ parallel lines

☐ perpendicular lines

☐ Pythagorean Theorem

☐ remote interior angles

☐ transversal

☐ triangle

☐ vertex

Are You Ready?

Study the Quick Review to see if you are ready to start this module.
Then complete the Quick Check.

Quick Review

Example 1
Solve equations.

Solve $34 + a + 68 = 180$.

$34 + a + 68 =$	180	Write the equation.
$102 + a =$	180	Add 34 and 68.
-102	$= -102$	Subtraction Property of Equality
$a = 78$		Simplify.

Example 2
Graph ordered pairs.

Graph $A(-1, 3)$, $B(2, 1)$, and $C(0, -4)$.

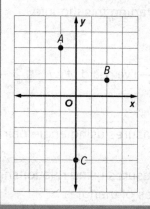

Start at the origin. The first number is the x-coordinate and the second number is the y-coordinate.

Quick Check

1. The equation $180 - x = 132$ represents the number of stamps in Colin's collection. Solve the equation for x.

2. Use the coordinate plane in Example 2 to graph and label each point on the coordinate plane.

   a. $X(-2, 2)$ **b.** $Y(3, 3)$ **c.** $Z(-3, -5)$

How Did You Do?

Which exercises did you answer correctly in the Quick Check?
Shade those exercise numbers at the right.

Angle Relationships and Parallel Lines

I Can... use the relationships between angles to find the measures of missing angles.

Explore Parallel Lines and Transversals

Online Activity You will use Web Sketchpad to explore the relationships between angles created by parallel lines and transversals.

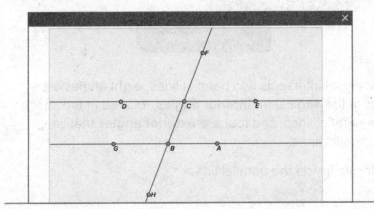

Learn Lines, Angles, and Transversals

Pairs of angles can be classified by their relationship to each other. A special case occurs when two lines intersect in a plane to form a right angle. These lines are **perpendicular lines**. Special notation is used to indicate perpendicular lines. Read $\ell \perp m$ as *line ℓ is perpendicular to line m.*

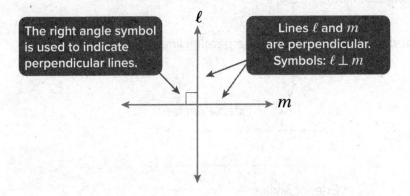

The right angle symbol is used to indicate perpendicular lines.

Lines ℓ and m are perpendicular.
Symbols: $\ell \perp m$

(continued on next page)

What Vocabulary Will You Learn?

alternate exterior angles

alternate interior angles

corresponding angles

exterior angles

interior angles

parallel lines

perpendicular lines

transversal

If the transversal is perpendicular to one of the parallel lines, what relationship does the transversal have to the other parallel line?

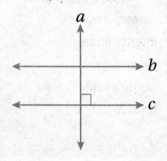

Two lines in a plane that never intersect are called **parallel lines**. A line that intersects two or more other lines in a plane is called a **transversal**. Special notation is used to indicate parallel lines. Read $s \parallel t$ as *line s is parallel to line t*.

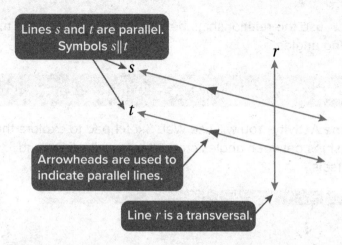

Lines s and t are parallel. Symbols $s \parallel t$

Arrowheads are used to indicate parallel lines.

Line r is a transversal.

When a transversal intersects two parallel lines, eight angles are formed. Four of the angles are **interior angles**, located in the space between the parallel lines, and four are **exterior angles** that lie outside the parallel lines.

Interior angles lie inside the parallel lines.

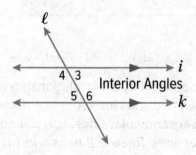

Interior Angles

Examples: $\angle 3$, $\angle 4$, $\angle \boxed{}$, $\angle \boxed{}$

Exterior angles lie outside the parallel lines.

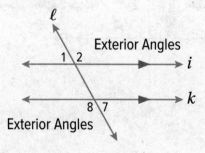

Exterior Angles

Exterior Angles

Examples: $\angle 1$, $\angle 2$, $\angle \boxed{}$, $\angle \boxed{}$

(continued on next page)

When two parallel lines are cut by a transversal, there is a relationship between the angles that are created. The angles in certain angle pairs, **alternate interior angles**, **alternate exterior angles**, and **corresponding angles**, have the same angle measure.
Special notation is used to indicate the measure of an angle.
Read $m\angle 1$ as *the measure of angle 1*.

Alternate interior angles are interior angles that lie on opposite sides of the transversal. When the lines are parallel, their measures are equal.

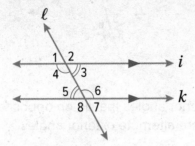

Examples: $m\angle 4 = m\angle 6$ and $m\angle 3 = m\angle$ ☐

Alternate exterior angles are exterior angles that lie on opposite sides of the transversal. When the lines are parallel, their measures are equal.

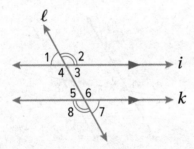

Examples: $m\angle 1 = m\angle 7$ and $m\angle 2 = m\angle$ ☐

Corresponding angles are those angles that are in the same position on the two lines in relation to the transversal. When the lines are parallel, their measures are equal.

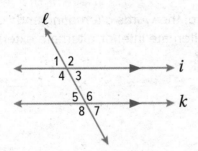

Examples: $m\angle 1 = m\angle 5$, $m\angle 2 = m\angle 6$, $m\angle 3 = m\angle 7$, and $m\angle 4 = m\angle$ ☐

Example 1 Classify Angle Pairs

Example 1 Classify Angle Pairs

Classify the relationship between ∠1 and ∠7 in the figure as *alternate interior*, *alternate exterior*, or *corresponding*.

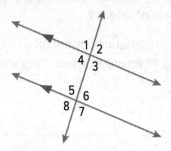

∠1 and ∠7 are exterior angles that lie on opposite sides of the transversal. They are alternate exterior angles.

Check

In the figure, the two lines shown are parallel and intersected by a transversal. Classify the relationship between ∠2 and ∠4 as *alternate exterior*, *alternate interior*, or *corresponding*.

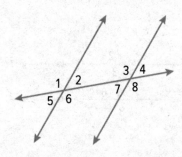

 **Go Online** You can complete an Extra Example online.

Pause and Reflect

How can the meaning of the words *alternating* and *corresponding* help you think about alternate interior, alternate exterior, and corresponding angles?

> Record your observations here

🍥 Think About It!

What are the locations of the angles with respect to the parallel lines?

💬 Talk About It!

Name another pair of alternate exterior angles. How many pairs of alternate exterior angles are there when two parallel lines are cut by a transversal? Will this happen when any two parallel lines are cut by a transversal? Explain.

Example 2 Classify Angle Pairs

Classify the relationship between ∠2 and ∠6 in the figure as _alternate interior_, _alternate exterior_, or _corresponding_.

🔵 **Think About It!**

What are the locations of the angles with respect to the parallel lines?

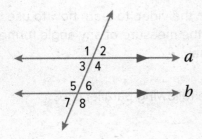

∠2 and ∠6 are in the _____ position on the two lines in relation to the transversal. They are corresponding angles.

Check

In the figure, the two lines shown are parallel and intersected by a transversal. Classify the relationship between ∠4 and ∠5 as _alternate exterior_, _alternate interior_, or _corresponding_.

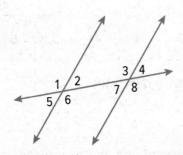

💬 **Talk About It!**

Name another pair of corresponding angles. How many pairs of corresponding angles are there when two parallel lines are cut by a transversal? Will this happen when any two parallel lines are cut by a transversal? Explain.

🔘 **Go Online** You can complete an Extra Example online.

Pause and Reflect

How did what you know about _alternate interior_, _alternate exterior_, and _corresponding angles_ help you solve the problem?

Record your observations here

Learn Find Missing Angle Measures

When two parallel lines are cut by a transversal, eight angles are formed. Special relationships exist among pairs of angles.

Go Online Watch the video to learn how to use these relationships to find the measure of any angle formed by two parallel lines and a transversal.

The video shows the following parallel lines.

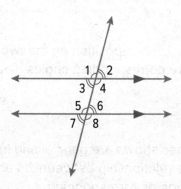

Complete the missing angle measures in the table.

Angle	1	2	3	4	5	6	7	8
Measure	105°	75°	75°	105°				

If you know the measure of one angle, you can use your knowledge of supplementary and vertical angles to find the measures of the three angles that are along the same line.

In the figure below, suppose $m\angle 1 = 50°$.

$m\angle 2 = 130°$ because $\angle 1$ and $\angle 2$ are _____.

$m\angle 3 = 50°$ because $\angle 1$ and $\angle 3$ are _____ angles.

$m\angle 4 = 130°$ because $\angle 1$ and $\angle 4$ are _____.

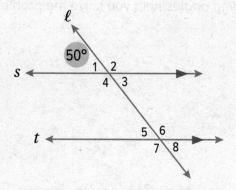

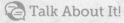

Talk About It!

Once you know the measures of angles 2, 3, and 4, how can you find the measures of angles 5, 6, 7, and 8?

Example 3 Find Missing Angle Measures

Mrs. Kumar designed the bookcase shown.
Line *a* is parallel to line *b*.

**If m∠2 = 105°, find m∠6 and m∠3. Justify
your answer.**

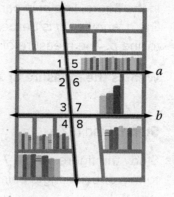

Part A Find m∠6.

Since ∠2 and ∠6 are supplementary, the
sum of their measures is _____.

So, m∠6 = 180° − 105° or 75°.

Part B Find m∠3.

Angle 6 and ∠3 are interior angles that lie on opposite sides of the
transversal. Since they are alternate interior angles, their measures
are _____.

So, m∠3 = 75°.

Check

Arianna's house has the porch stairs shown. Line *m* is parallel to
line *n*. If m∠7 = 35°, find m∠1 and m∠2.

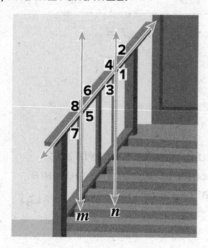

Show
your work
here

Go Online You can complete an Extra Example online.

🗨 Think About It!

Think about the
special relationship
between ∠2 and ∠6.
How does ∠3 relate to
either ∠2 or ∠6?

🗨 Talk About It!

How many unique
angle measurements
exist in the figure?

Example 4 Find Missing Angle Measures

In the figure, line *m* is parallel to
line *n*, and line *q* is perpendicular
to line *p*. The measure of ∠1 is 40°.

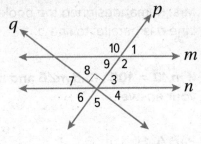

What is the measure of ∠7?

Think About It!

If you know *m*∠8, and
want to find the
measure of ∠7, what
other angle will help
you?

Step 1 Find *m*∠6.

Study the figure. Angle 7 is adjacent to angle 6 and angles 1 and 6
form a special angle pair. Find *m*∠6 first. Then use *m*∠6 to find *m*∠7.

Since ∠1 and ∠6 are alternate exterior angles, their measures
are equal. The *m*∠1 is 40°, so the *m*∠6 = 40°.

Talk About It!

If *m*∠1 = 40°, do you
have enough
information to find all
of the missing angles in
the figure? Explain.

Step 2 Find *m*∠7.

Since ∠6, ∠7, and ∠8 form a straight line, the sum of their measures
is 180°.

$$m\angle6 + m\angle7 + m\angle8 = 180°$$ Write the equation.

$$\boxed{} + m\angle7 + \boxed{} = 180°$$ Replace *m*∠6 with 40°
and *m*∠8 with 90°.

$$\boxed{} + m\angle7 = 180°$$ Add.

$$\underline{-130° \qquad\qquad = -130°}$$ Subtraction Property
of Equality

$$m\angle7 = \boxed{}$$ Simplify.

So, *m*∠7 is 50°.

Check

In the figure, line *a* is parallel to line *b*, and
line *c* is perpendicular to line *d*. The measure
of ∠7 is 125°. What is the measure of ∠4?

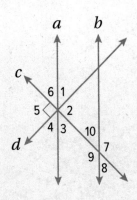

🅡 **Go Online** You can complete an Extra Example online.

🌐 Apply Construction

In the photo of the bridge, line a is parallel to line b. If $m\angle 1 = 16x°$ and $m\angle 2 = (10x + 30)°$, find $m\angle 3$.

1 What is the task?

Make sure you understand exactly what question to answer or problem to solve. You may want to read the problem three times. Discuss these questions with a partner.

First Time Describe the context of the problem, in your own words.
Second Time What mathematics do you see in the problem?
Third Time What are you wondering about?

2 How can you approach the task? What strategies can you use?

3 What is your solution?

Use your strategy to solve the problem.

4 How can you show your solution is reasonable?

🪃 **Write About It!** Write an argument that can be used to defend your solution.

💬 Talk About It!

What is the relationship between angles 1 and 2? How does that help you solve the problem?

Check

In the painting, line *a* is parallel to line *b*. The measure of angle 1 is $(5x + 24)°$ and the measure of angle 2 is $7x°$. Find $m\angle 1$.

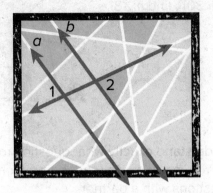

Show your work here

Go Online You can complete an Extra Example online.

Pause and Reflect

What have you learned about the angles formed by *parallel lines* and *transversals*? Can you name the angles that are formed? Can you determine which angles have the same measure?

Record your observations here

Practice

Go Online You can complete your homework online.

For Exercises 1–4, use the figure at the right. In the figure, line *m* is parallel to line *n*. For each pair of angles, classify the relationship in the figure as *alternate interior*, *alternate exterior*, or *corresponding*. (Examples 1 and 2)

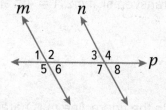

1. ∠2 and ∠7

2. ∠1 and ∠3

3. ∠4 and ∠5

4. ∠5 and ∠7

5. Arturo is designing a bridge for science class using parallel supports for the top and bottom beam. Find $m\angle 2$ and $m\angle 3$ if $m\angle 1 = 60°$. Justify your answer. (Example 3)

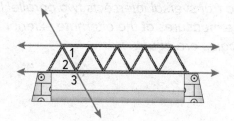

6. In the figure, line *m* is parallel to line *n*. The measure of ∠3 is 58°. What is the measure of ∠7? (Example 4)

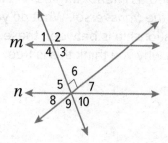

7. The symbol below is an equal sign with a slash through it. It is used to represent *not equal to* in math, as in $x \neq 5$. If $m\angle 1 = 108°$, classify the relationship between ∠1 and ∠2. Then find $m\angle 2$. Assume the equal sign consists of parallel lines.

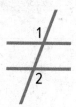

Test Practice

8. **Multiselect** In the figure, line *m* and line *n* are parallel. Select all of the statements that are true.

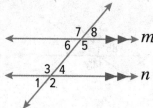

☐ ∠1 and ∠8 are alternate exterior angles.

☐ ∠3 and ∠7 are corresponding angles.

☐ ∠2 and ∠8 are corresponding angles.

☐ ∠4 and ∠6 are alternate interior angles.

☐ ∠5 and ∠7 are corresponding angles.

9. Angles A and B are corresponding angles formed by two parallel lines cut by a transversal. If $m\angle A = 4x°$ and $m\angle B = (3x + 7)°$, find the value of x. Explain.

10. In the figure, line m is parallel to line n. If $m\angle 3 = (7x − 10)°$ and $m\angle 6 = (5x + 10)°$, what are the measures of $\angle 3$ and $\angle 6$?

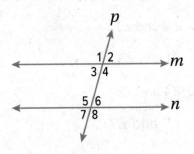

11. **MP Reason Abstractly** Refer to the figure in Exercise 10. Look at a pair of angles described as *interior angles on the same side of the transversal*. What do you think the relationship is between these angles? Explain why you think this is true.

12. Determine if the statement is *true* or *false*. Construct an argument that can be used to defend your solution.

If a transversal intersects two parallel lines, the measures of the alternate exterior angles are equal.

13. Determine the measure of $\angle W$. Construct an argument that can be used to defend your solution.

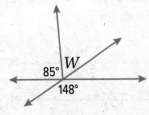

14. **MP Find the Error** A student was finding the measure of $\angle 5$ in the figure below. She concluded that $m\angle 5 = 86°$ because it is a corresponding angle with $\angle 2$. Find her mistake and correct it.

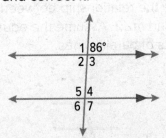

Angle Relationships and Triangles

I Can... find the measures of interior and exterior angles in a triangle by using relationships between these angles.

What Vocabulary Will You Learn?

exterior angle

interior angles

line segment

remote interior angles

triangle

vertex

Learn Triangles

A **line segment** is part of a line containing two endpoints and all of the points between them. A **triangle** is formed by three line segments that intersect at their endpoints. A point where the segments intersect is a **vertex**. The three angles that lie inside a triangle, formed by the segments and the vertices, are called **interior angles**.

Triangle XYZ, written $\triangle XYZ$, has sides and angles that can be named using its vertices X, Y, and Z. The angle located at vertex Y can be named with symbols as $\angle Y$, $\angle XYZ$, or $\angle ZYX$. The sides of a triangle can be named using segment notation. For example, $\overline{XY}$ is read as *segment XY*. Name the missing sides, vertices, and angles.

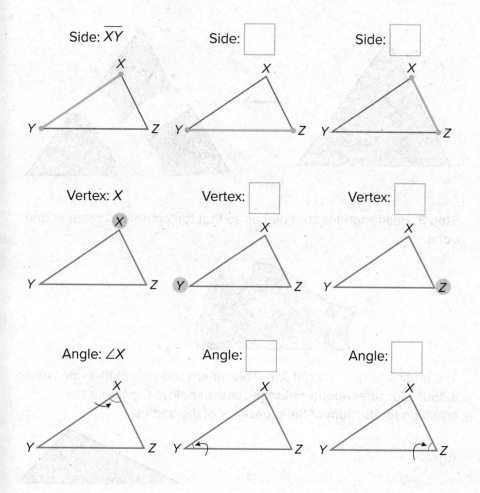

Side: $\overline{XY}$

Side: ☐

Side: ☐

Vertex: X

Vertex: ☐

Vertex: ☐

Angle: $\angle X$

Angle: ☐

Angle: ☐

Explore Angles of Triangles

Online Activity You will use Web Sketchpad to explore the relationship among the angle measures in triangles.

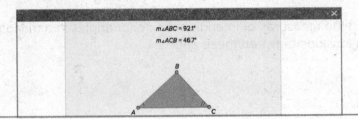

Learn Angle Sum of Triangles

The measures of the angles in a triangle have a special relationship.

Go Online Watch the video and follow these steps to learn about the relationship among the angles in a triangle.

Step 1 Draw a triangle like the one shown below.

Step 2 Tear off each corner.

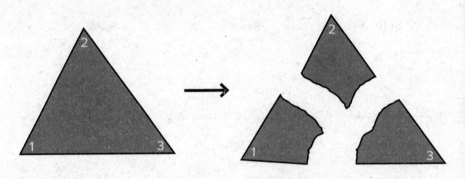

Step 3 Rearrange the torn pieces so that the corners all meet at one point.

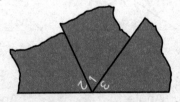

The angles form a straight line. This means the sum of their measures is 180°. Consider another triangle, shown below. Complete the equation for the sum of the measures of the angles.

$110° +$ ⬚ $+$ ⬚ $=$ ⬚

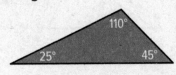

(continued on next page)

The activity on the previous page illustrates the relationship among the angle measures of a triangle.

Words	Model
The sum of the measures of the interior angles of a triangle is 180°.	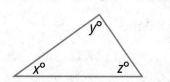
Variables	
$x + y + z = 180$	

🌐 **Example 1** Find Missing Angle Measures

Find the value of *x* in the flag of Saint Kitts and Nevis.

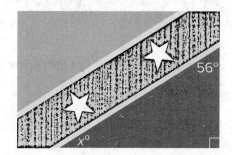

$x + 56 + 90 = 180$ Write the equation.

$x + \boxed{} = 180$ Add.

$\underline{-146 = -146}$ Subtraction Property of Equality

$x = \boxed{}$ Simplify.

So, the value of *x* in the triangle is 34.

Check

What is the value of *x* in the doghouse shown?

Show your work here

🌐 **Go Online** You can complete an Extra Example online.

Example 2 Use Ratios to Find Angle Measures

In $\triangle ABC$, the measures of the angles A, B, and C, respectively, are in the ratio 1:4:5.

Find the measure of each angle.

Step 1 Write an equation.

Words
The sum of the angle measurements in a triangle is 180°.
Variable
Let x represent the measure of angle A.
The measure of angle B is 4 times greater, or $4x$. The measure of angle C is 5 times greater than x, or $5x$.
Equation
$x + 4x + 5x = 180$

Step 2 Solve the equation and evaluate the angle measurements.

$$x + 4x + 5x = 180 \qquad \text{Write the equation.}$$
$$10x = 180 \qquad \text{Combine like terms.}$$
$$x = \boxed{} \qquad \text{Simplify.}$$

Since $x = 18$, $m\angle A$ is 18°. The measure of $\angle B$ is $4x°$, or $4\left(\boxed{}\right)$, which is 72°. The measure of $\angle C$ is $5x°$, or $5\left(\boxed{}\right)$, which is 90°.

Check

In $\triangle LMN$, the measures of the angles L, M, and N, respectively, are in the ratio 1:2:5. Find the measure of each angle.

Show your work here

Go Online You can complete an Extra Example online.

Explore Exterior Angles of Triangles

▶ **Online Activity** You will use Web Sketchpad to explore the relationship between an exterior angle and two remote interior angles of a triangle.

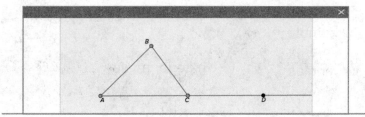

Learn Exterior Angles of Triangles

In addition to its three interior angles, a triangle can have an **exterior angle** formed by one side of the triangle and the extension of the adjacent side.

In the diagram shown, angles 4, 5, and 6 are exterior angles. An exterior angle is supplementary to its adjacent interior angle because the two angles form a straight line.

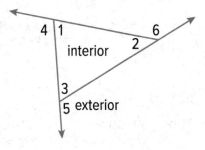

Complete the statements relating each exterior angle and its adjacent interior angle.

$m\angle 4 + m\angle 1 = 180°$

$m\angle 6 + m\angle \boxed{} = 180°$

$m\angle 5 + m\angle \boxed{} = 180°$

Each exterior angle of the triangle has two **remote interior angles** that are *not* adjacent to the exterior angle. Angle 4 is an exterior angle of the triangle. Its two remote interior angles are $\angle 2$ and $\angle 3$.

Which angles are remote interior angles in relation to $\angle 5$? $\angle \boxed{}$ and $\angle \boxed{}$

Which angles are remote interior angles in relation to $\angle 6$?

$\angle \boxed{}$ and $\angle \boxed{}$

(continued on next page)

Go Online Watch the video to learn about the relationship between an exterior angle of a triangle and its two remote interior angles.

The video demonstrates this relationship using triangle *ABC*.

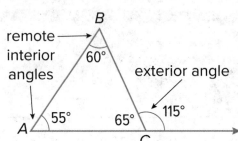

$60° + 55° + 65° =$ [ ]　　　　Angle sum of a triangle

$65° + 115° =$ [ ]　　　　　　Supplementary angles

$60° + 55° + 65° = 65° + 115°$　　Write the equation.

$\underline{\qquad - 65° = - 65° \qquad}$　　Subtract 65 from each side.

[ ] $+$ [ ] $=$ [ ]　　　　　　Simplify.

The equation shows that the sum of the measures of $\angle A$ and $\angle B$ is equal to the measure of the exterior angle.

Talk About It!

Can the measure of an exterior angle be less than or equal to either of its remote interior angles?

Words
The measure of an exterior angle of a triangle is equal to the sum of the measures of its two remote interior angles.
Model
  A   B   1   C
Symbols
$m\angle A + m\angle B = m\angle 1$

🌐 Example 3 Find Exterior Angle Measures

In the beach chair shown,
$m\angle 2 = 55°$ and $m\angle 3 = 60°$.

Find the measure of ∠1.

💭 Think About It!

What steps do you need to take to find $m\angle 1$?

Angle 1 is an exterior angle. Its two remote interior angles are ∠2 and ∠3.

$m\angle 2 + m\angle 3 = m\angle 1$ Write the equation.

☐ + ☐ $= m\angle 1$ $m\angle 2 = 55°, m\angle 3 = 60°$

☐ $= m\angle 1$ Simplify.

So, the measure of ∠1 is 115°.

💬 Talk About It!

What is another way to find the measure of angle 1?

Check

In the roof frame shown, $m\angle 2 = 25°$ and $m\angle 3 = 45°$. Find the measure of ∠1.

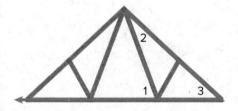

Show your work here

🅑 **Go Online** You can complete an Extra Example online.

Example 4 Use Exterior Angles to Find Missing Angle Measures

In the figure, $m\angle 4 = 135°$.

Find the measures of $\angle 2$ and $\angle 1$.

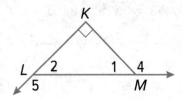

Angle 4 is an exterior angle. Its two remote interior angles are $\angle 2$ and $\angle LKM$.

$$m\angle 2 + m\angle LKM = m\angle 4 \qquad \text{Write the equation.}$$

$$m\angle 2 + \boxed{} = \boxed{} \qquad m\angle LKM = 90° \text{ and } m\angle 4 = 135°$$

$$m\angle 2 = \boxed{} \qquad \text{Subtraction Property of Equality}$$

Since $\angle 1$ and $\angle 4$ are supplementary, the sum of their measures is

_____.

So, $m\angle 1$ is $180° - 135°$ or $45°$.

Check

In the figure, $m\angle 5 = 147°$. Find the measures of $\angle 1$ and $\angle 2$.

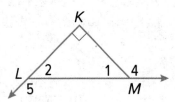

Show
your work
here

Go Online You can complete an Extra Example online.

Apply Geometry

What are the measures of ∠CDE and ∠BCE in the figure?

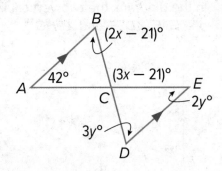

1 What is the task?

Make sure you understand exactly what question to answer or problem to solve. You may want to read the problem three times. Discuss these questions with a partner.

First Time Describe the context of the problem, in your own words.
Second Time What mathematics do you see in the problem?
Third Time What are you wondering about?

2 How can you approach the task? What strategies can you use?

3 What is your solution?

Use your strategy to solve the problem.

 Talk About It!

Is there more than one way to find the measure of ∠BCE? Explain.

4 How can you show your solution is reasonable?

✎ **Write About It!** Write an argument that can be used to defend your solution.

Check

In the diagram, the two vertical lines are parallel. Find the measures of ∠FDE, ∠DEF, and ∠FHG.

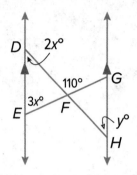

Show your work here

Go Online You can complete an Extra Example online.

Pause and Reflect

Explain what you have learned about the interior angles, exterior angles, and remote interior angles of a triangle. Given a drawing, can you identify these angles?

Record your observations here

Practice

➤ **Go Online** You can complete your homework online.

Find the value of x in each object. (Example 1)

1.

2.

3. In △FGH, the measures of angles F, G, and H, respectively, are in the ratio 4:4:10. Find the measure of each angle. (Example 2)

4. In the knitting pattern, m∠1 = 42°. Find the measure of ∠2. (Example 3)

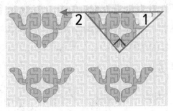

Test Practice

5. In the figure, m∠4 = 74° and m∠3 = 43°. Find the measures of ∠1 and ∠2. (Example 4)

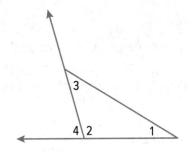

6. Open Response What is the measure of ∠x, in degrees, in the figure shown?

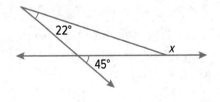

Apply

7. What are the measures of ∠ADC and ∠DCB in the figure below?

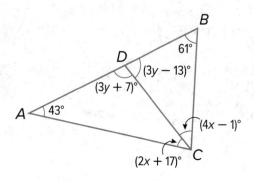

8. What are the measures of ∠CAB and ∠ACB in the figure below?

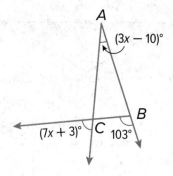

9. (MP) **Find the Error** A student is finding the measures of the angles in a triangle that have the ratio 4:4:7. Find the mistake and correct it.

$$4x + 4x + 7x = 180$$

$$15x = 180$$

$$x = 12$$

So, the angle measures are 12, 12, and 84.

10. (MP) **Persevere with Problems** The measure of ∠A in △ABC is twice the measure of ∠B, and ∠C is 20° less than the measure of ∠B. What are the measures of the angles in △ABC?

11. Determine if the statement is *true* or *false*. Construct an argument that can be used to defend your solution.

An exterior angle of a triangle will always be obtuse.

12. (MP) **Find the Error** A student states that the exterior angle of a triangle can never be a right angle. Find the mistake and correct it.

The Pythagorean Theorem

I Can... find the measures of the sides of a right triangle using the Pythagorean Theorem and square roots.

Learn Right Triangles

A right triangle is a triangle with one right angle. The **legs** are the sides that form the right angle.

The **hypotenuse** is the side opposite the right angle. It is the longest side of the triangle.

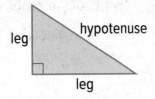

Explore Right Triangle Relationships

Online Activity You will use Web Sketchpad to explore the relationship among the sides in right triangles.

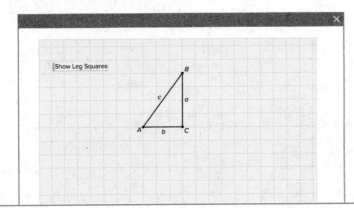

Pause and Reflect

Where did you encounter struggle in this Explore, and how did you deal with it? Write down any questions you still have.

Record your observations here

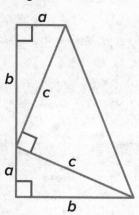
Learn The Pythagorean Theorem

The **Pythagorean Theorem** describes the relationship between the lengths of the legs and the length of the hypotenuse for any right triangle.

Words	Model
In a right triangle, the sum of the squares of the lengths of the legs is equal to the square of the length of the hypotenuse.	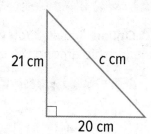

Symbols	
$a^2 + b^2 = c^2$	

Go Online Watch the animation to see how you can use the Pythagorean Theorem to find the length of the hypotenuse of a right triangle when you know the length of the two legs.

The animation shows how to find the missing length in the right triangle shown using the following steps.

$$a^2 + b^2 = c^2 \qquad \text{Pythagorean Theorem}$$

$$21^2 + 20^2 = c^2 \qquad \text{Replace } a \text{ with 21 and } b \text{ with 20.}$$

$$\boxed{} + \boxed{} = c^2 \qquad \text{Evaluate.}$$

$$\boxed{} = c^2 \qquad \text{Add.}$$

$$\pm\sqrt{841} = c \qquad \text{Definition of square root}$$

$$\pm \boxed{} = c \qquad \text{Simplify.}$$

The equation has two solutions. However, the length of any side of a triangle cannot be negative. So, the hypotenuse has a length of 29 centimeters.

🌐 **Example 1** Find the Hypotenuse

A ladder is leaning against a building as shown in the figure.

What is the length of the ladder?

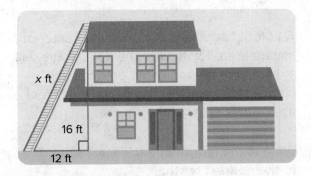

x ft

16 ft

12 ft

💭 Think About It!

What side of the triangle is the hypotenuse? How does its length compare to the lengths of the other two sides?

You know the lengths of the legs. Use the Pythagorean Theorem to find the hypotenuse. The legs are represented by a and b, and it does not matter which leg is represented by which letter.

$$a^2 + b^2 = c^2 \qquad \text{Pythagorean Theorem}$$

$$\boxed{}^2 + \boxed{}^2 = c^2 \qquad \text{Replace } a \text{ with 12 and } b \text{ with 16.}$$

$$\boxed{} + \boxed{} = c^2 \qquad \text{Evaluate.}$$

$$\boxed{} = c^2 \qquad \text{Add.}$$

$$\pm\sqrt{400} = c \qquad \text{Definition of square root}$$

$$\pm\boxed{} = c \qquad \text{Simplify.}$$

The equation has two solutions. However, the length of a side must be positive. So, the ladder is 20 feet long.

💬 Talk About It!

The sides of the triangle are in the ratio 12:16:20. Simplify this ratio. How can you use this ratio to verify the length of the hypotenuse is correct?

Check

The mast of a sailboat is 15 feet tall. A support rope is attached to the top of the mast and attaches to the deck 8 feet away from the base of the mast. Find the length, x, of the support rope.

Show your work here

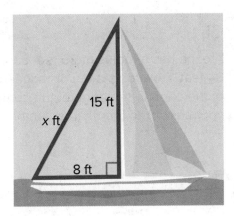

15 ft

x ft

8 ft

🅝 **Go Online** You can complete an Extra Example online.

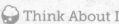
What is the relationship between the legs of the triangle and the hypotenuse?

⊕ Example 2 Find the Hypotenuse in Three Dimensions

A 12-foot flagpole is placed in the center of a square area. To stabilize the pole, a wire will stretch from the top of the pole to each corner of the square. The flagpole is 7 feet from each corner of the square.

What is the length of each wire? Round to the nearest tenth.

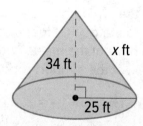

The legs are represented by line segments *AB* and *AC*. The hypotenuse is represented by line segment *BC*. Use the Pythagorean Theorem to find the length of the wire, *BC*.

$$AB^2 + AC^2 = BC^2 \qquad \text{Pythagorean Theorem}$$

$$\boxed{}^2 + \boxed{}^2 = BC^2 \qquad AB = 7 \text{ and } AC = 12.$$

$$\boxed{} + \boxed{} = BC^2 \qquad \text{Evaluate.}$$

$$\boxed{} = BC^2 \qquad \text{Add.}$$

$$\pm\sqrt{193} = BC \qquad \text{Definition of square root}$$

$$\pm \boxed{} \approx BC \qquad \text{Simplify.}$$

Since the length cannot be negative, the length of the wire is about 13.9 feet.

Talk About It!

The wire will stretch from the top of the pole to each corner of the square. How can you determine the approximate total length of wire that is needed to stabilize the pole?

Check

A cone has a radius of 25 feet and a height of 34 feet. How long is the distance from the top of the cone to the edge of the cone? Round your answer to the nearest tenth.

Show your work here

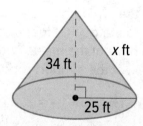

> ⓘ **Go Online** You can complete an Extra Example online.

🌐 Example 3 Find Missing Leg Lengths

A plane takes off from an airport and travels 13 miles on its path.

If the plane is 12 miles from its takeoff point horizontally, what is its height?

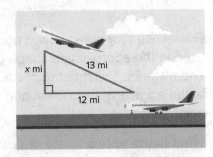

💭 **Think About It!**

Which side is the hypotenuse and which sides are the legs?

The distance between the plane's location and the takeoff point is the hypotenuse of a right triangle.

$a^2 + b^2 = c^2$ Pythagorean Theorem

$\boxed{}^2 + x^2 = \boxed{}^2$ Replace the variables with side lengths.

$\boxed{} + x^2 = \boxed{}$ Evaluate.

$x^2 = \boxed{}$ Subtraction Property of Equality

$x = \pm\sqrt{25}$ Definition of square root

$x = \pm\boxed{}$ Simplify.

Since length cannot be negative, the height of the plane is 5 miles.

💬 **Talk About It!**

Explain why it makes sense that the plane's path during takeoff covers a greater distance than the horizontal distance along the ground.

Check

Julia let out 50 meters of kite string when she notices that her kite is directly above her friend Sasha. If Julia is 35 meters from Sasha, what is the height of the kite? Round to the nearest tenth.

Show your work here

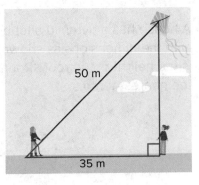

🌐 **Go Online** You can complete an Extra Example online.

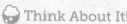
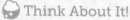 Talk About It!

Suppose the height and radius of the cone are doubled to 16 feet and 30 feet, respectively. What effect does this have on the length of the hypotenuse? Explain your reasoning.

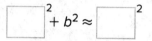 **Example 4** Find Missing Leg Lengths in Three Dimensions

Gravel used for construction purposes is piled in the approximate shape of a cone. The distance from the top of the cone to the edge is about 17 feet and the height is about 8 feet.

What is the approximate radius of the cone? Round to the nearest tenth if necessary.

The slant height, height, and radius form a right triangle.

$$a^2 + b^2 = c^2 \qquad \text{Pythagorean Theorem}$$

$$\boxed{}^2 + b^2 \approx \boxed{}^2 \qquad \text{Replace the variables with known lengths.}$$

$$\boxed{} + b^2 \approx \boxed{} \qquad \text{Evaluate.}$$

$$b^2 \approx \boxed{} \qquad \text{Subtraction Property of Equality}$$

$$b \approx \pm\sqrt{225} \qquad \text{Definition of square root}$$

$$b \approx \pm \boxed{} \qquad \text{Simplify.}$$

Length cannot be negative, so the radius of the cone is about 15 feet.

Check

A house has a pyramid-shaped roof with the dimensions as shown. What is the height of the roof? Round to the nearest tenth.

 Show your work here

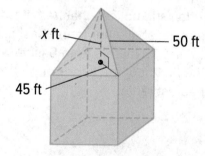

Go Online You can complete an Extra Example online.

🌐 **Apply** Maps

Alma has a motor boat that averages 3 miles per gallon of gasoline, and the tank holds 15 gallons of gasoline. At 9 a.m., Alma left the dock. At 10 a.m., her position was 3 miles west and 4 miles north of the dock. If she continues at this rate, in how many more hours will the tank be out of gasoline?

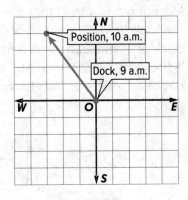

🎬 **Go Online** Watch the animation.

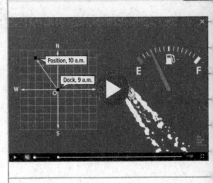

1 What is the task?

Make sure you understand exactly what question to answer or problem to solve. You may want to read the problem three times. Discuss these questions with a partner.

First Time Describe the context of the problem, in your own words.
Second Time What mathematics do you see in the problem?
Third Time What are you wondering about?

2 How can you approach the task? What strategies can you use?

Record your observations here

3 What is your solution?

Use your strategy to solve the problem.

Show your work here

💬 Talk About It!

After how many more hours should she turn around in order to make it back to the dock?

4 How can you show your solution is reasonable?

🔺 **Write About It!** Write an argument that can be used to defend your solution.

Check

Meredith has a boat that averages 5 miles per gallon of gasoline and the tank holds 17 gallons of gasoline. At 3 p.m., Meredith left the dock. At 4 p.m., she was 5 miles east and 12 miles south of the dock. If she continues at this rate, how many more hours will the tank be out of gasoline? Round to the nearest tenth if necessary.

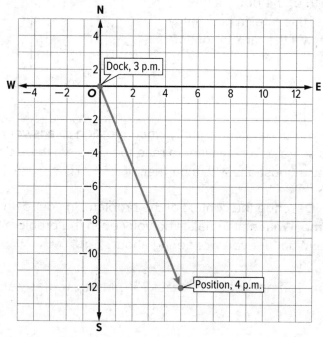

Show your work here

Go Online You can complete an Extra Example online.

Pause and Reflect

Where did you encounter struggle in the Apply? Did you understand the information given? Did you understand what you were asked to find? Study the example to develop mastery of your area(s) of struggle.

Record your observations here

Learn Geometric Proof

A *proof* is a logical argument where each statement is justified by a reason. Proofs are used in geometry to explain or prove why something is true about a geometric figure.

Use the following steps to write a proof.

Step 1: Given Information

List the given information, or what you know. If possible, draw a diagram to illustrate the information.

Step 2: Hypothesis

State what is to be proven.

Step 3: Statements

Create an argument by forming a logical chain of statements linking the given information to what you are trying to prove.

Step 4: Reasons

Justify each statement with a reason. Reasons include definitions, algebraic properties, and theorems.

Step 5: Prove (Conclusion)

State what it is you have proven.

Explore Proof of the Pythagorean Theorem

Online Activity You will use Web Sketchpad to explore proving the Pythagorean Theorem.

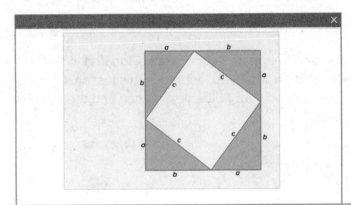

Pause and Reflect

Think about how the five steps listed in the Learn function to prove that something is true. For each step, write a sentence explaining why that step is important.

Record your observations here

Foldables It's time to update your Foldable, located in the Module Review, based on what you learned in this lesson. If you haven't already assembled your Foldable, you can find the instructions on page FL1.

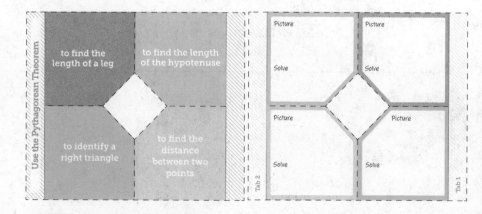

Name _____ Period _____ Date _____

Practice

Go Online You can complete your homework online.

1. What is the length of a diagonal of a rectangular picture whose sides are 12 inches by 17 inches? Round to the nearest tenth. **(Example 1)**

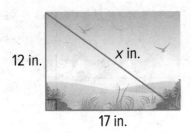

12 in.

x in.

17 in.

2. How far is the airplane from the runway? Round to the nearest tenth. **(Example 2)**

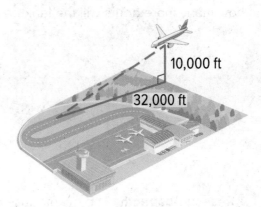

10,000 ft

32,000 ft

3. The diagonal of a television measures 27 inches. If the width is 22 inches, calculate its height to the nearest inch. **(Example 3)**

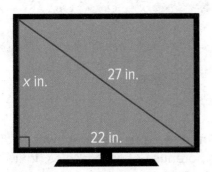

x in.

27 in.

22 in.

4. The distance from the top of the cone to the edge is 15 feet. The height of the cone is 6 feet. What is the radius of the cone? Round to the nearest tenth. **(Example 4)**

6 ft

15 ft

5. What is the perimeter of a right triangle if the hypotenuse is 15 centimeters and one of the legs is 9 centimeters?

Test Practice

6. Multiselect Select all of the following statements that are true about the right triangle shown.

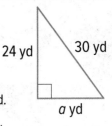

24 yd

30 yd

a yd

☐ The hypotenuse is 30 yd.

☐ The missing leg is 18 yd.

☐ The missing leg is 24 yd.

☐ The formula $24^2 + a^2 = 30^2$ can be used to find the missing leg measure.

☐ The formula $30^2 + a^2 = 24^2$ can be used to find the missing leg measure.

Apply

7. Patrick has a gas-powered boat that averages 5 miles per gallon of gasoline, and the tank holds 4 gallons. At 2 p.m., he left the dock. At 3 p.m., he was 6 miles west and 8 miles south of the dock. If he continues at this rate, in how many more hours will the tank be out of gasoline?

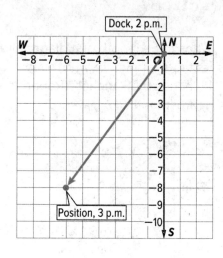

8. **MP Reason Abstractly** The hypotenuse of a right triangle is 54 inches long. Find possible measures for the legs of the triangle. Round to the nearest tenth. Write an argument that can be used to defend your solution.

9. **MP Persevere with Problems** The radius of the circle shown is 4 centimeters. One of the legs of the right triangle formed is also 4 centimeters. Find the length of leg x. Round to the nearest tenth.

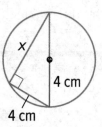

10. April wants to find the distance from point A to point G in the figure. Give two measurements she will need to find this distance.

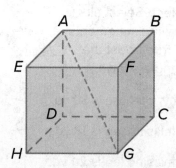

11. **MP Find the Error** A student was finding the hypotenuse of an isosceles right triangle. The legs were each 20 millimeters. Find the mistake and correct it.

$$(20 + 20)^2 = c^2$$
$$40^2 = c^2$$
$$40 = c$$

Converse of the Pythagorean Theorem

I Can... determine if a triangle is a right triangle by using the converse of the Pythagorean Theorem.

What Vocabulary Will You Learn?

converse

converse of the Pythagorean Theorem

Learn Converse of the Pythagorean Theorem

The **converse of the Pythagorean Theorem** states that if the sides of a triangle have lengths a, b, and c units such that $a^2 + b^2 = c^2$, then the triangle is a right triangle.

If you reverse the *if* and *then* statements of the Pythagorean Theorem, you have formed its **converse**.

Statement: If a triangle is a right triangle, then $a^2 + b^2 = c^2$.

Converse: If $a^2 + b^2 = c^2$, then the triangle is a right triangle.

The converse of the Pythagorean Theorem is also true.

Consider the triangle, with dimensions 3 meters, 4 meters, and 5 meters, as shown. To determine if the triangle is a right triangle, use the converse of the Pythagorean Theorem.

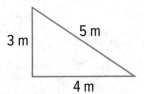

3 m   5 m   4 m

If $a^2 + b^2 = c^2$, then the triangle is a right triangle.

$$a^2 + b^2 = c^2 \qquad \text{Pythagorean Theorem}$$

$$3^2 + 4^2 \stackrel{?}{=} 5^2 \qquad \text{Replace } a \text{ with 3, } b \text{ with 4, and } c \text{ with 5.}$$

$$9 + 16 \stackrel{?}{=} 25 \qquad \text{Evaluate.}$$

$$25 = 25 \qquad \text{Add.}$$

Since the side lengths satisfy the converse of the Pythagorean Theorem, the triangle is a right triangle.

💭 **Think About It!**

How can you use the converse of the Pythagorean Theorem to solve this problem?

💬 **Talk About It!**

Suppose a triangle has side lengths that are half as long as the side lengths of the triangle shown. Is the new triangle a right triangle also? Explain.

🌐 **Example 1** Use the Converse of the Pythagorean Theorem

A carpenter is framing a wall and needs to ensure that the wooden boards form a right angle.

Determine whether the triangle shown is a right triangle.

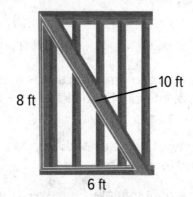

If $a^2 + b^2 = c^2$, then the triangle is a right triangle. The lengths of the sides of the triangle are 6 feet, 8 feet, and 10 feet.

$$a^2 + b^2 \overset{?}{=} c^2 \qquad \text{Pythagorean Theorem}$$

$$\boxed{}^2 + \boxed{}^2 \overset{?}{=} \boxed{}^2 \qquad \text{Replace } a \text{ with 6, } b \text{ with 8, and } c \text{ with 10.}$$

$$\boxed{} + \boxed{} \overset{?}{=} \boxed{} \qquad \text{Evaluate.}$$

$$\boxed{} = 100 \qquad \text{Add.}$$

Since the side lengths do satisfy the converse of the Pythagorean Theorem, the triangle is a right triangle.

Check

A construction worker is measuring the frame for a roof to ensure that it creates a right triangle. Determine whether the triangle is a right triangle.

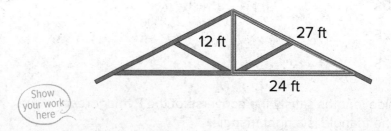

Show your work here

🌐 **Go Online** You can complete an Extra Example online.

⊕ **Example 2** Use the Converse of the Pythagorean Theorem

Angela is creating a rectangular patio in her back yard. In order to ensure that the patio space has right-angle corners, she measures the diagonal to create a triangle.

Determine whether the triangle shown is a right triangle.

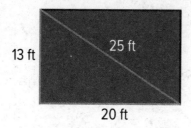

13 ft 25 ft 20 ft

$a^2 + b^2 \overset{?}{=} c^2$ Pythagorean Theorem

$\boxed{}^2 + \boxed{}^2 \overset{?}{=} \boxed{}^2$ Replace a with 13, b with 20, and c with 25.

$\boxed{} + \boxed{} \overset{?}{=} \boxed{}$ Evaluate.

$\boxed{} \neq 625$ Add.

Since the side lengths do not satisfy the converse of the Pythagorean Theorem, the triangle is not a right triangle.

Check

Four stakes mark the foundation of a house. In order to make sure the foundation is rectangular, the diagonal is also marked. Determine whether the triangle is a right triangle.

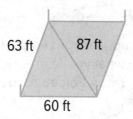

63 ft 87 ft 60 ft

Show your work here

 Go Online You can complete an Extra Example online.

Explore Prove the Converse of the Pythagorean Theorem

Online Activity You will use Web Sketchpad to explore proving the converse of the Pythagorean Theorem.

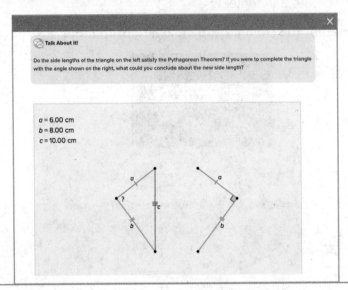

Talk About It!

Do the side lengths of the triangle on the left satisfy the Pythagorean Theorem? If you were to complete the triangle with the angle shown on the right, what could you conclude about the new side length?

a = 6.00 cm
b = 8.00 cm
c = 10.00 cm

Foldables It's time to update your Foldable, located in the Module Review, based on what you learned in this lesson. If you haven't already assembled your Foldable, you can find the instructions on page FL1.

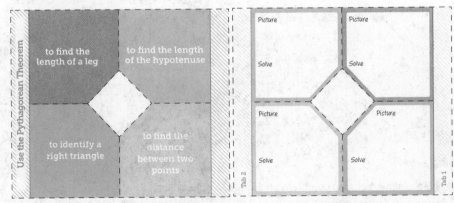

Practice

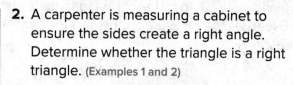 Go Online You can complete your homework online.

1. Three cities form a triangle. Tom measures the distances between the three cities on a map. The distances between the three cities are 45 miles, 56 miles, and 72 miles. Is the triangle formed by the three cities a right triangle? (Examples 1 and 2)

2. A carpenter is measuring a cabinet to ensure the sides create a right angle. Determine whether the triangle is a right triangle. (Examples 1 and 2)

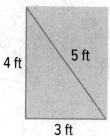

4 ft   5 ft

3 ft

3. Allie wants to make sure that the pieces of cloth for a costume are right triangles. Determine whether the triangle is a right triangle. (Examples 1 and 2)

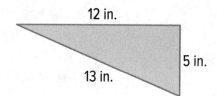

12 in.

5 in.

13 in.

4. In order to ensure that the roof consists of right angles, an architect measures the diagonal to create a triangle. If the dimensions of the triangle are 9.5 feet, 16 feet, and 18.5 feet, is the triangle a right triangle? (Examples 1 and 2)

5. Elyse is building a square raised-bed garden with 4-foot sides. She measured the diagonal to be $4\sqrt{2}$ feet. Is her garden square? Explain.

6. The distance between each base on a softball field is 60 feet. Maddie is placing the bases and measures the distance between home plate and second base. She determines the distance to be 80 feet. Are the bases at right angles? Explain.

Test Practice

7. Three islands form a triangle. Island A is $7\frac{1}{2}$ miles from Island B and 18 miles from Island C. Island B is $19\frac{1}{2}$ miles from Island C. Is the triangle formed a right triangle? Explain.

8. **Multiselect** Select all of the following that *cannot* be the measures of sides of a right triangle.

   ☐ 6 cm, 8 cm, 10 cm

   ☐ 14 cm, 18 cm, 20 cm

   ☐ 20 cm, 21 cm, 29 cm

   ☐ 6 cm, 6 cm, 6 cm

   ☐ 10 cm, 24 cm, 26 cm

9. Manuel measured the distance from the top vertex of the triangle shown to its base. He found the distance to be 5 feet. Did he measure the height? Explain your response.

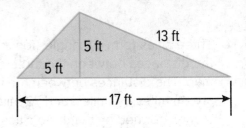

10. **Reason Abstractly** The measurements 3, 4, and 5 satisfy the converse of the Pythagorean Theorem and therefore form a right triangle. If each measurement is doubled, will the new triangle still be a right triangle? Write an argument that can be used to defend your solution.

11. Explain how you can use the converse of the Pythagorean Theorem to determine if a parallelogram is a rectangle.

12. When using the Pythagorean Theorem in real-world problems, why is it sometimes necessary to round to a specific decimal place?

13. **Find the Error** A student knows that a triangle with sides 8 inches, 15 inches, and 17 inches is a right triangle. He concludes that if he adds 2 inches to each side, the triangle will still be a right triangle. Find his error and correct it.

Distance on the Coordinate Plane

I Can... find the distance between two points on a coordinate plane using the Pythagorean Theorem.

Explore Use the Pythagorean Theorem to Find Distance

▶ **Online Activity** You will use Web Sketchpad to explore how to find distance on the coordinate plane using the Pythagorean Theorem.

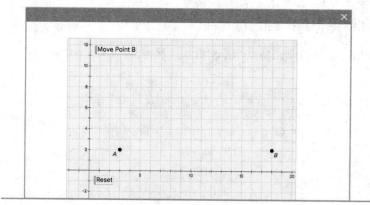

Learn Find Distance on the Coordinate Plane

▶ **Go Online** Watch the video to see how to use the Pythagorean Theorem to find the distance between (3, 1) and (7, −2) on the coordinate plane.

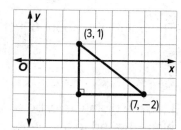

Step 1 Use the points to draw a right triangle.
The segment between the two points is the hypotenuse of the right triangle.

Step 2 Determine the lengths of the legs by counting units. The vertical leg has a length of 3 units. The horizontal leg has a length of 4 units.

Step 3 Determine the length of the hypotenuse.

$$a^2 + b^2 = c^2$$ Pythagorean Theorem
$$3^2 + 4^2 = c^2$$ Replace a with 3 and b with 4.
$$25 = c^2$$ Simplify.
$$5 = c$$ Find the positive square root of 25.

The points are 5 units apart.

💬 Talk About It!
Is there a way to determine the length of each leg in the right triangle that can be formed by only examining the coordinates of the points? Explain your reasoning.

How can you use the Pythagorean Theorem to find the distance between the two points?

Study the coordinates for the points graphed. How can you determine that the side lengths are 4 and 5 just by looking at the coordinates?

Example 1 Find Distance on the Coordinate Plane

Find the distance, c, between (3, 0) and (7, −5) on the coordinate plane. Round to the nearest tenth.

Step 1 Graph the points on the coordinate plane.

Plot the points, (3, 0) and (7, −5), on the coordinate plane and connect with a segment. This segment will be the hypotenuse of the right triangle. Then draw two other segments to form a right triangle. Determine the length of each leg. The lengths of the legs are _____ units and _____ units.

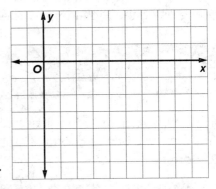

Step 2 Use the Pythagorean Theorem.

Find the length of the hypotenuse, which is the distance between the two points.

$$a^2 + b^2 = c^2 \qquad \text{Pythagorean Theorem}$$

$$\boxed{}^2 + \boxed{}^2 = c^2 \qquad \text{Replace } a \text{ with 4 and } b \text{ with 5.}$$

$$\boxed{} = c^2 \qquad \text{Evaluate.}$$

$$\pm\sqrt{41} = c \qquad \text{Definition of square root}$$

$$\pm \boxed{} \approx c \qquad \text{Use a calculator.}$$

The points are about 6.4 units apart.

Check

Find the distance, c, between (−3, −4) and (1, 2) on the coordinate plane. Round to the nearest tenth.

(Show your work here)

🅑 **Go Online** You can complete an Extra Example online.

🌐 Apply Maps

The map shows two walking trails in a park. Both paths end at the picnic area, located at $C(6, 1)$. Trail A starts at $A(-6, 6)$, and Trail B starts at $B(-10, -11)$. If each unit on the map represents 0.5 mile, how many miles longer is Trail B than Trail A?

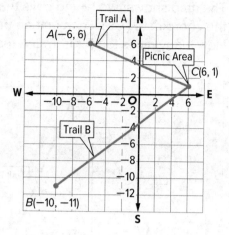

1 What is the task?

Make sure you understand exactly what question to answer or problem to solve. You may want to read the problem three times. Discuss these questions with a partner.

First Time Describe the context of the problem, in your own words.
Second Time What mathematics do you see in the problem?
Third Time What are you wondering about?

2 How can you approach the task? What strategies can you use?

3 What is your solution?

Use your strategy to solve the problem.

🗨 **Talk About It!**

Suppose each unit on the map represented 1.5 miles. How would that change your answer?

4 How can you show your solution is reasonable?

🖉 **Write About It!** Write an argument that can be used to defend your solution.

Check

The map shows two biking trails that meet at a park, located at the coordinate (1, 9). Each unit on the map represents 0.5 mile. How many miles longer is Trail A than Trail B?

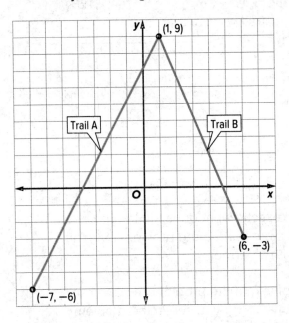

Show your work here

Go Online You can complete an Extra Example online.

Foldables It's time to update your Foldable, located in the Module Review, based on what you learned in this lesson. If you haven't already assembled your Foldable, you can find the instructions on page FL1.

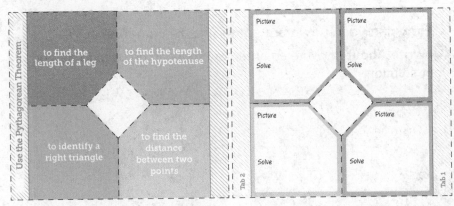

Practice

🞂 Go Online You can complete your homework online.

Find the distance, c, between each pair of points on the coordinate plane.
Round to the nearest tenth if necessary. (Example 1)

1. (−4, −3), (2, 1)

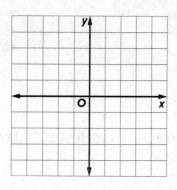

2. (0, 2), (5, −2)

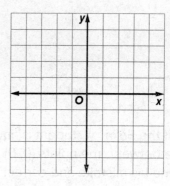

3. (0, 0), (−4, −3)

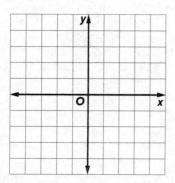

4. (−3, 4), (2, −3)

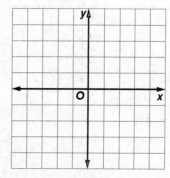

Test Practice

5. An archaeologist at a dig sets up a coordinate system using string. Two similar artifacts are found—one at position (1, 4) and the other at (5, 2). How far apart were the two artifacts? Round to the nearest tenth of a unit if necessary.

6. Equation Editor The coordinates of points A and B are (−7, 5) and (4, −3), respectively. What is the distance, in units, between the points? Round to the nearest tenth.

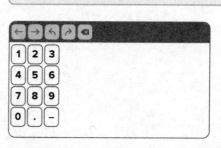

7. Rosa is looking at a map of the park that is laid out on a coordinate plane. Rosa is at $(1, -1)$. The shelter house is at $(-2, -4)$ and the fossil exhibit is at $(3, 2)$. Each unit on the map represents 100 feet. How much closer is Rosa to the fossil exhibit than to the shelter house?

8. (MP) **Find the Error** A student is finding the distance between the points $(0, 0)$ and $(4, 5)$. Find his mistake and correct it.

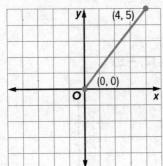

$4^2 + 5^2 = 16 + 25$
$\qquad = 41$

So, the distance is 41 units.

9. Find the area of square *ABCD* shown below.

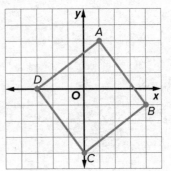

10. Name the coordinates of the endpoints of a line segment that is neither horizontal nor vertical and has a length of 5 units.

11. (MP) **Reason Abstractly** Determine if the following statement is *true* or *false*. Write an argument that can be used to defend your solution.

When finding the distance between two points on a coordinate plane, it is always necessary to use the Pythagorean Theorem.

📖 **Foldables** Use your Foldable to help review the module.

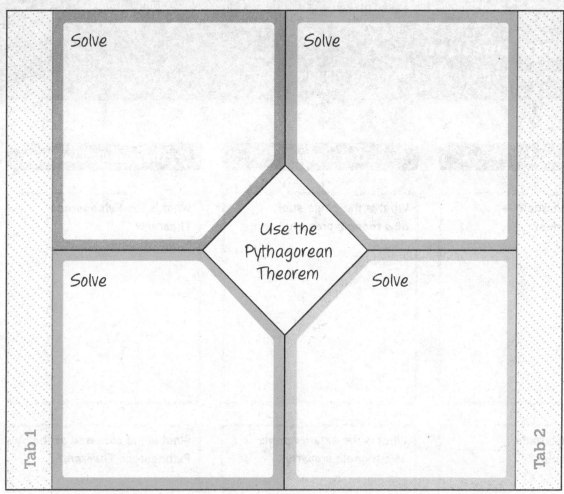

Solve

Solve

Use the Pythagorean Theorem

Solve

Solve

Tab 1

Tab 2

Rate Yourself! ⬛ ◆ ★

Complete the chart at the beginning of the module by placing a checkmark in each row that corresponds with how much you know about each topic after completing this module.

Write about one thing you learned.	Write about a question you still have.

Reflect on the Module

Use what you learned about triangles and the Pythagorean Theorem to complete the graphic organizer.

e Essential Question

How can angle relationships and right triangles be used to solve everyday problems?

Parallel Lines

What are alternate interior angles?

What are alternate exterior angles?

What are corresponding angles?

Triangles

What is the angle sum of a triangle property?

What is the exterior angle of a triangle property?

How are these properties useful?

Pythagorean Theorem

What is the Pythagorean Theorem?

What is the converse of the Pythagorean Theorem?

How is it useful?

Test Practice

1. Multiple Choice Which of the following pairs of angles are *alternate interior angles?* (Lesson 1)

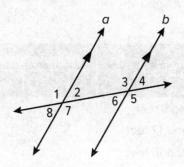

(A) ∠2 and ∠4

(B) ∠2 and ∠6

(C) ∠1 and ∠5

(D) ∠4 and ∠5

2. Equation Editor In the figure, line *a* is parallel to line *b*, and line *c* is perpendicular to line *d*. The measure of ∠10 is 140°. What is the measure, in degrees, of ∠4? (Lesson 1)

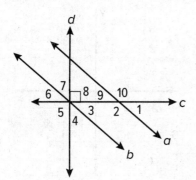

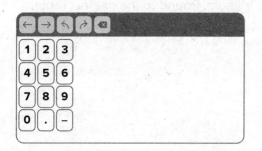

3. Multiselect In △JKL, the measures of the angles J, K, and L, respectively, are in the ratio 3:3:6. Which of the following statements are accurate regarding the angle measures? Select all that apply. (Lesson 2)

☐ Angles J and K have equal measures.

☐ The measure of ∠L is half the measure of ∠J.

☐ $m\angle L = 45°$

☐ $m\angle K = 90°$

☐ The angles form an isosceles right triangle.

☐ The measure of ∠L is twice the measure of ∠K.

4. Open Response In the stained glass shown, suppose $m\angle 1 = 47°$. Find the measure of ∠2. Show your work. (Lesson 2)

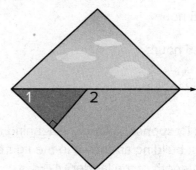

5. Multiple Choice Suki has a boat that averages 6 miles per gallon of gasoline, and the tank holds 8 gallons of gasoline. At 2 p.m., Suki left the dock at the marina. At 3 p.m., she was 8 miles east and 15 miles north of the dock. (Lesson 3)

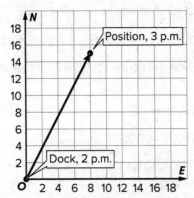

If she continues at this rate, which of the following represents how many more hours until the tank will be out of gasoline (rounded to the nearest tenth)?

(A) 1.5 hours

(B) 1.8 hours

(C) 2.5 hours

(D) 2.8 hours

6. Open Response A ladder is leaning against a brick building as shown in the figure. What is the length of the ladder? (Lesson 3)

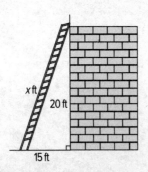

7. Table Item Jack is building a ramp. He wants to ensure that the side view of the ramp is in the shape of a right triangle. Indicate which of the sets of dimensions form a right triangle. (Lesson 4)

Ramp Dimensions	Right Triangle?	
	yes	no
3, 4, and 5 feet		
5, 10, and 12 feet		
9, 12, and 15 feet		
10, 15, and 20 feet		

8. Grid Find the distance, c, between $(-2, 1)$ and $(4, 3)$ on the coordinate plane. Round to the nearest tenth. (Lesson 5)

A. Graph and connect the points on the coordinate plane to create a segment to be used as a hypotenuse. Then use the hypotenuse to create the legs of a right triangle.

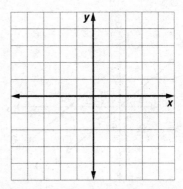

B. Use the Pythagorean Theorem to find the length of the hypotenuse of your right triangle. Round to the nearest tenth.

Module 8
Transformations

e Essential Question
What does it mean to perform a transformation on a figure?

What Will You Learn?
Place a checkmark (✓) in each row that corresponds with how much you already know about each topic **before** starting this module.

KEY			Before			After		
⬛ — I don't know. ◗ — I've heard of it. ★ — I know it!			⬛	◗	★	⬛	◗	★
translating figures on the coordinate plane								
using coordinate notation to describe translations								
reflecting figures on the coordinate plane								
using coordinate notation to describe reflections								
rotating figures on the coordinate plane								
using coordinate notation to describe rotations								
dilating figures on the coordinate plane								
using coordinate notation to describe dilations								

📒 Foldables Cut out the Foldable and tape it to the Module Review at the end of the module. You can use the Foldable throughout the module as you learn about transformations.

What Vocabulary Will You Learn?

Check the box next to each vocabulary term that you may already know.

☐ center of dilation   ☐ line of reflection   ☐ scale factor

☐ center of rotation   ☐ preimage   ☐ transformation

☐ dilation   ☐ reflection   ☐ translation

☐ image   ☐ rotation

Are You Ready?

Study the Quick Review to see if you are ready to start this module.
Then complete the Quick Check.

Quick Review

Example 1

Graph polygons on a coordinate plane.

Two vertices of a rectangle are $A(-3, 4)$ and $B(1, 4)$. One side is 5 units. Graph rectangle $ABCD$ and label the other two vertices.

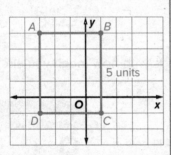

Example 2

Add integers.

Find $3 + (-8)$.

$3 + (-8) = -5$ $|3| - |-8| = -5$
The sum is negative because $|-8| > |3|$.

Quick Check

1. Two vertices of a square are $M(-2, 3)$ and $N(2, 3)$. One side is 4 units. Graph square $MNPQ$ and label the other two vertices.

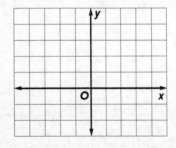

2. A fish was 6 meters below sea level. The fish descended 19 meters. Find $-6 + (-19)$ to determine the location of the fish compared to sea level.

How Did You Do?

Which exercises did you answer correctly in the Quick Check?
Shade those exercise numbers at the right.

Translations

I Can... translate figures on the coordinate plane and use coordinate notation to describe translations.

Learn Transformations

A **transformation** is an operation that maps an original geometric figure onto a new figure. A transformation can slide, flip, turn, or resize a figure.

The original geometric figure is called a **preimage**, and the new figure is called the **image.**

The graph shows a point on the coordinate plane that has been transformed to a new point.

On the graph, A is the preimage and A′ is the image. A′ is read "A prime." Prime symbols are used for vertices in a transformed image.

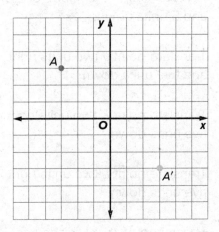

Learn Translations on a Coordinate Plane

A **translation** is a transformation that slides a figure from one position to another without turning it.

When translating a figure, every point of the preimage is moved the same distance and in the same direction.

The triangle shown is translated 3 units to the left and 2 units down.

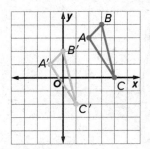

The image and the preimage are congruent.

What Vocabulary Will You Learn?
image

preimage

transformation

translation

💬 Talk About It!

Describe the image if the line shown is translated 2 units up.

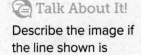

How would you begin
translating △JKL?

Example 1 Translate Figures on the Coordinate Plane

The graph of △JKL is shown.

Graph the image of △JKL after a translation of 2 units right and 5 units down. Write the coordinates of the image.

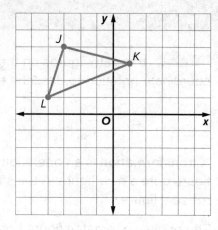

Part A Graph the image of △JKL after a translation of 2 units right and 5 units down.

Translate point J two units right and five units down.

Translate point K two units right and five units down.

Translate point L two units right and five units down.

The image of △JKL is shown.

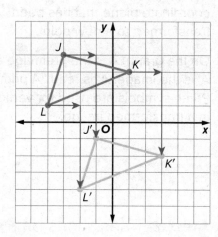

Part B Write the coordinates of the image.

Use the graph to write the coordinates of the vertices of the image.

$J(-3, 4) \rightarrow J'(\boxed{}, \boxed{})$

$K(1, 3) \rightarrow K'(\boxed{}, \boxed{})$

$L(-4, 1) \rightarrow L'(\boxed{}, \boxed{})$

Talk About It!

How do the x-values of the coordinates of the preimage and image compare? the y-values?

Check

The graph of △*DEF* has coordinates *D*(1, 1), *E*(3, −2), and *F*(4, 3). Graph △*DEF* and its image after a translation of 5 units left and 3 units down. Write the coordinates of the image.

Part A   Graph △*DEF* and its image after a translation of 5 units left and 3 units down.

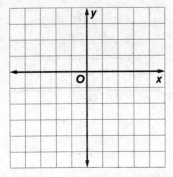

Part B   Write the coordinates of the image.

Go Online You can complete an Extra Example online.

Explore Translate Using Coordinates

Online Activity You will use Web Sketchpad to explore how to translate figures using coordinates.

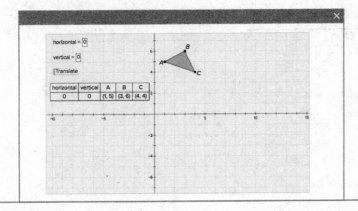

Learn Translations Using Coordinates

The coordinates of a translated image can be determined using coordinate notation.

Go Online Watch the animation to learn about coordinate notation for a translation.

The animation shows that you can use the following coordinate notation to describe the translation from J to J'.

$(x, y) \rightarrow (x + a, y + b)$

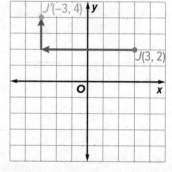

The value of a is the number of units the pre-image is translated left or right. The value of b is the number of units the pre-image is translated up or down. Since point J moved _____ units to the left and _____ units up, the translation described using coordinate notation is $(x, y) \rightarrow (x - 6, y + 2)$.

Words
Horizontal Translation of a The image's x-coordinate is found by adding a to the preimage's x-coordinate.
Vertical Translation of b The image's y-coordinate is found by adding b to the preimage's y-coordinate.

Model	Symbols
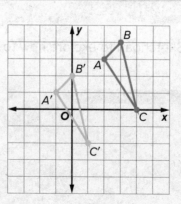	$(x, y) \rightarrow (x + a, y + b)$
	Example
	$(x, y) \rightarrow (x - 3, y - 2)$

Example 2 Translate Using Coordinates

Triangle *XYZ* has vertices *X*(−1, −2), *Y*(6, −3), and *Z*(2, −5).

Write the coordinate notation for a translation of 2 units left and 1 unit up. Then, write the coordinates of △*X′Y′Z′*.

Part A Write the coordinate notation for a translation of 2 units left and 1 unit up.

To translate the triangle 2 units left, _____ 2 from each *x*-coordinate. To translate the triangle 1 units up, _____ 1 to each *y*-coordinate.

So, the coordinate notation is $(x, y) \rightarrow (x - 2, y + 1)$.

Part B Write the coordinates of △*X′Y′Z′*.

Vertices of △*XYZ*	$(x, y) \rightarrow (x - 2, y + 1)$	Vertices of △*X′Y′Z′*
X(−1, −2)	(−1 − 2, −2 + 1)	*X′*(☐ , ☐)
Y(6, −3)	(6 − 2, −3 + 1)	*Y′*(☐ , ☐)
Z(2, −5)	(2 − 2, −5 + 1)	*Z′*(☐ , ☐)

So, the vertices of *X′Y′Z′* are *X′*(−3, −1), *Y′*(4, −2), and *Z′*(0, −4).

Check

Triangle *ABC* has vertices *A*(3, 2), *B*(1, −3), and *C*(−5, 0). Write the coordinate notation for a translation of 6 units left and 9 units down. Then, write the coordinates of △*A′B′C′*.

Part A Write the translation in coordinate notation.

Part B Write the coordinates of the image.

Show your work here

 Think About It!

Do you think it's necessary to graph the pre-image and image to solve this problem? Why or why not?

Go Online You can complete an Extra Example online.

Think About It!

What is coordinate notation for a translation?

Talk About It!

How can this translation be described in words?

Talk About It!

How can you make sure that the coordinate notation in Example 3 is correct?

Example 3 Use Coordinate Notation to Describe Translations

Use coordinate notation to describe the translation.

Choose a point on the preimage and its corresponding point on the image. For example, Point A is located at $(-7, 6)$. Point A' is located at $(2, -1)$.

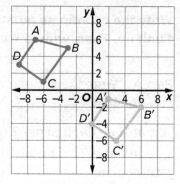

$(x, y) \rightarrow (x + a, y + b)$ Write the coordinate notation.

$(-7, 6) \rightarrow ($ ☐ $+ a,$ ☐ $+ b)$ Replace x and y with the coordinates of Point A.

The coordinates of the image are represented by $(-7 + a, 6 + b)$. You need to find the values of a and b. Since you know the coordinates of Point $A'(2, -1)$, you can write two equations to find the values of a and b.

$-7 + a = 2$ $6 + b = -1$ Write equations to solve for a and b.

$a = $ ☐ $b = $ ☐   Solve.

So, the translation described using coordinate notation is $(x, y) \rightarrow (x + 9, y - 7)$.

Check

Use coordinate notation to describe the translation.

Show your work here

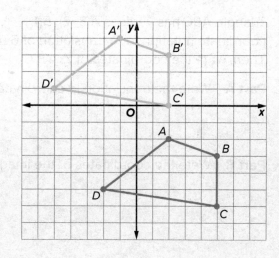

Go Online You can complete an Extra Example online.

🌐 Apply Map Reading

Emilia's house is located on the map shown. She walks 3 units east and 2 units north to meet up with a friend. She then walks 1 unit west and 3 units north to get to school. If she were to walk a straight path from her house to the school, what would be the distance? Round the answer to the nearest tenth.

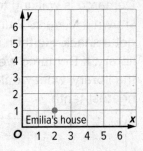

Emilia's house

1 What is the task?

Make sure you understand exactly what question to answer or problem to solve. You may want to read the problem three times. Discuss these questions with a partner.

First Time Describe the context of the problem, in your own words.
Second Time What mathematics do you see in the problem?
Third Time What are you wondering about?

2 How can you approach the task? What strategies can you use?

Record your observations here

3 What is your solution?

Use your strategy to solve the problem.

Show your work here

4 How can you show your solution is reasonable?

✏️ **Write About It!** Write an argument that can be used to defend your solution.

💬 Talk About It!

How did understanding translations help you solve the problem?

Check

Demarco is walking to the library. His house is located on the map shown. He walks 2 units west and 4 units south to stop at the store to get a snack. He then walks 4 units west and 1 unit north to get to the library. If Demarco were to walk in a straight path from his house to the library, what would be the distance? Round your answer to the nearest tenth.

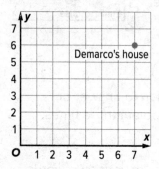

Show your work here

Go Online You can complete an Extra Example online.

📖 **Foldables** It's time to update your Foldable, located in the Module Review, based on what you learned in this lesson. If you haven't already assembled your Foldable, you can find the instructions on page FL1.

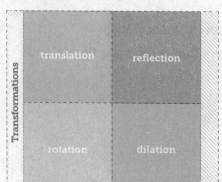

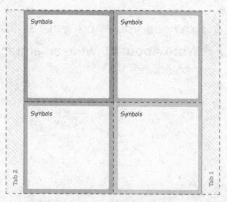

Practice

1. The graph of △ABC is shown. Graph the image of △ABC after a translation of 4 units right and 1 unit up. Write the coordinates of the image. (Example 1)

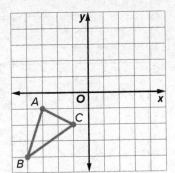

2. The graph of △EFG is shown. Graph the image of △EFG after a translation of 3 units left and 1 unit down. Write the coordinates of the image. (Example 1)

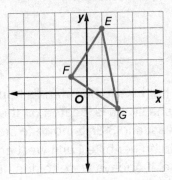

Triangle QRS has vertices Q(−2, 2), R(−3, −4), and S(1, −2). Write the coordinate notation for each translation given. Then write the coordinates of △Q'R'S' after each translation. (Example 2)

3. 7 units right and 4 units down

4. 2 units left and 3 units up

5. The preimage and image of WXYZ are shown. Use coordinate notation to describe the translation. (Example 3)

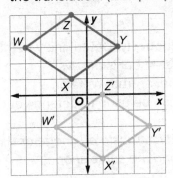

Test Practice

6. **Open Response** Triangle JKL has vertices J(−2, 2), K(−3, −4), and L(1, −2). Write the coordinate notation for a translation of 8 units right and 1 unit up.

7. Jabari is walking to the park from school. The school is located on the map shown. He walks 2 units west and 4 units south to stop at a store to buy a snack. He then walks 6 units west and 4 units south to get to the park. If Jabari were to walk in a straight path from the school to the park, what would be the distance? Round the answer to the nearest tenth.

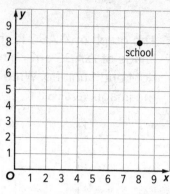

8. Andaya is riding her bike to her friend's house from swim practice. The pool where she swam is located on the map shown. She rides her bike 5 units east and 3 units north to stop at her house. She then continues riding her bike 1 unit west and 2 units north to get to her friend's house. If Andaya were to ride her bike in a straight path from the pool to her friend's house, what would be the distance? Round the answer to the nearest tenth.

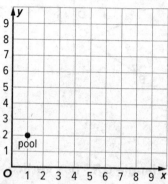

9. ⓂⓅ **Reason Inductively** A figure is translated by $(x, y) \rightarrow (x + 3, y - 4)$ then by $(x, y) \rightarrow (x - 3, y + 4)$. Without graphing, how do you know the final position of the figure? Write an argument that can be used to defend your solution.

10. ⓂⓅ **Identify Structure** A point is located at (x, y). Write the coordinate notation for a translation of a units right and b units down.

11. A classmate states that a two-dimensional figure could have each of its vertices translated in different ways and it would still be considered a translation. Explain to your classmate why this is incorrect.

12. ⓂⓅ **Reason Inductively** Determine whether the following statement is *always*, *sometimes*, or *never* true. Write an argument that can be used to defend your solution.
A preimage and its translated image are the same size and the same shape.

Reflections

I Can... describe reflections of figures on the coordinate plane using coordinates and coordinate notation.

Learn Reflections on a Coordinate Plane

A **reflection** is a mirror image of the original figure. It is the result of a transformation of a figure across a line called a **line of reflection.**

When reflecting a figure, each point of the preimage and its image are the same distance from the line of reflection. The image and the preimage are congruent.

You can reflect a figure across the x-axis or across the y-axis.

x-axis

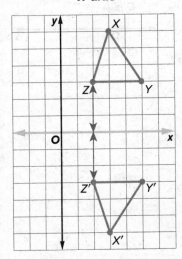

y-axis

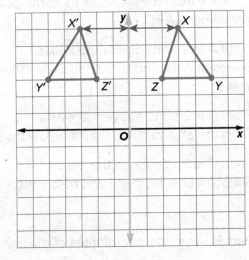

You can also reflect a figure across other lines. The reflection of △XYZ across the line y = 2 is shown below.

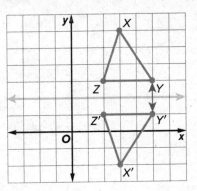

What Vocabulary Will You Learn?
line of reflection

reflection

💬 Talk About It!

Describe the image if the line shown is reflected across the x-axis.

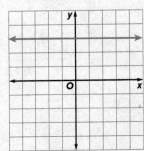

How far away from the
x-axis is each vertex of
the triangle?

Example 1 Reflect Figures on the Coordinate Plane

The graph of △*ABC* is shown.

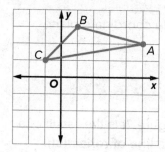

Graph the image of △*ABC* after a reflection across the *x*-axis. Write the coordinates of the image.

Part A Graph the image of △*ABC* after the reflection.

The *x*-axis is the line of reflection.
Plot each vertex of △*A'B'C'* the same
distance from the *x*-axis as its corresponding
vertex on △*ABC*.

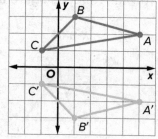

Point *A* is 2 units above the *x*-axis.
So, point *A'* is plotted 2 units below the
x-axis.

Point *B* is 3 units above the *x*-axis.
So, point *B'* is plotted 3 units below the *x*-axis.

Point *C* is 1 unit above the *x*-axis.
So, point *C'* is plotted 1 unit below the *x*-axis.

Part B Write the coordinates of the image.

Use the graph to write the coordinates of the vertices of the image.

$A(5, 2) \rightarrow A'(\boxed{}, \boxed{})$

$B(1, 3) \rightarrow B'(\boxed{}, \boxed{})$

$C(-1, 1) \rightarrow C'(\boxed{}, \boxed{})$

😮 Talk About It!

How do the *x*-values
of the coordinates of
the preimage and
image compare? the
y-values?

Check

The graph of $\triangle PQR$ has coordinates $P(1, 5)$, $Q(3, 7)$, and $R(4, -1)$. Graph $\triangle PQR$ and its image after a reflection across the y-axis. Then write the coordinates of the reflected image.

Part A Graph $\triangle PQR$ and its image after a reflection across the y-axis.

Part B Write the coordinates of the image.

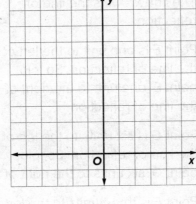

Show your work here

🐾 **Go Online** You can complete an Extra Example online.

Example 2 Reflect Figures on the Coordinate Plane

The graph of $\triangle EFG$ is shown.

Graph the image of $\triangle EFG$ after a reflection across the line $x = 3$. Write the coordinates of the image.

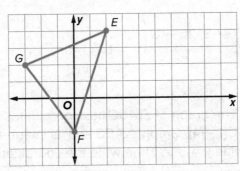

💭 **Think About It!**
What is true about the line of reflection?

Part A Graph the line of reflection. The line $x = 3$ is a vertical line, where each point on the line has an x-coordinate of 3.

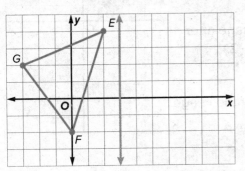

(continued on next page)

Part B Graph the image of △EFG after the reflection.

The line $x = 3$ is the line of reflection.

Point E is one unit to the left of the line of reflection. So, plot point E' one unit to the right of the line of reflection.

Point G is six units to the left of the line of reflection. So, plot point G' six units to the right of the line of reflection.

Point F is three units to the left of the line of reflection. So, plot point F' three units to the right of the line of reflection.

Then connect the points E', G', and F' to draw the image.

Talk About It!

How do the x-values of the coordinates of the preimage and image compare? the y-values? Will this happen every time a figure is reflected across a vertical line?

Part C Write the coordinates of the image.

Use the graph to write the coordinates of the vertices of the image.

$E(2, 4) \rightarrow E'\left(\boxed{}, \boxed{}\right)$

$F(0, -2) \rightarrow F'\left(\boxed{}, \boxed{}\right)$

$G(-3, 2) \rightarrow G'\left(\boxed{}, \boxed{}\right)$

Check

The graph of △ABC has coordinates $A(-3, -3)$, $B(-1, -5)$, and $C(1, -4)$. Graph △ABC and its image after a reflection across the line $y = -2$. Then write the coordinates of the image.

Part A Graph the line of reflection.

Part B Graph △ABC and its image after a reflection across the line $y = -2$.

Part C Write the coordinates of the image.

Go Online You can complete an Extra Example online.

Explore Reflect Using Coordinates

Online Activity You will use Web Sketchpad to explore how to reflect figures using coordinates.

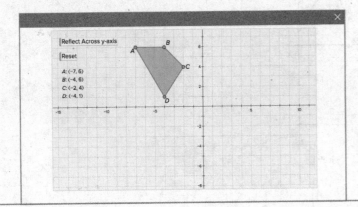

Learn Reflect Using Coordinates

Go Online Watch the animation to learn about coordinate notation for a reflection.

The animation shows $\triangle ABC$ and its image after a reflection across the x-axis. Write the coordinates of the image.

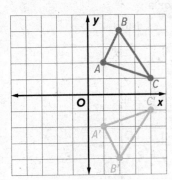

$A(1, 2) \rightarrow A'\left(1, \boxed{}\right)$

$B(2, 4) \rightarrow B'\left(2, \boxed{}\right)$

$C(4, 1) \rightarrow C'\left(4, \boxed{}\right)$

Notice that the y-coordinates of the image are the opposite of the y-coordinates of the preimage. So, the coordinate notation for a reflection across the x-axis is $(x, y) \rightarrow (x, -y)$.

(continued on next page)

The animation shows △ABC and its image after a reflection across the y-axis. Write the coordinates of the image.

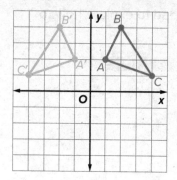

$A(1, 2) \rightarrow A'(\boxed{}, 2)$

$B(2, 4) \rightarrow B'(\boxed{}, 4)$

$C(4, 1) \rightarrow C'(\boxed{}, 1)$

Notice that the x-coordinates of the image are the opposite of the x-coordinates of the preimage. So, the coordinate notation for a reflection across the y-axis is $(x, y) \rightarrow (-x, y)$.

 Talk About It!

When a figure is reflected across the x-axis, explain why you can multiply the y-coordinate of the preimage by −1 to determine the y-coordinate of the image.

Reflections across the x-axis	
Words	**Model**
The x-coordinates are the same. The image's y-coordinates are opposites of those of the preimage's.	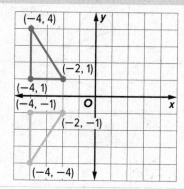
Symbols	
$(x, y) \rightarrow (x, -y)$	

Talk About It!

Which coordinate could you multiply by −1 when a figure is reflected across the y-axis?

Reflections across the y-axis	
Words	**Model**
The y-coordinates are the same. The image's x-coordinates are opposites of those of the preimage's.	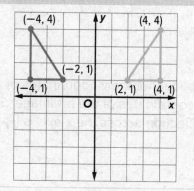
Symbols	
$(x, y) \rightarrow (-x, y)$	

Example 3 Reflect Using Coordinates

Triangle *RST* has coordinates *R*(2, 1), *S*(−1, 3), *T*(−2, 0). The triangle is reflected across the *x*-axis.

Write the coordinate notation for a reflection across the *x*-axis. Then write the coordinates of △*R'S'T'*.

Part A Write the coordinate notation for a reflection across the *x*-axis.

When a point is reflected across the *x*-axis, the _____ -coordinate remains the same. The _____ -coordinate of the image will be the opposite of the *y*-coordinate of the preimage.

So, the coordinate notation is $(x, y) \rightarrow (x, -y)$.

Part B Write the coordinates of △*R'S'T'*.

The *x*-coordinates remain the same. Write the opposite of each *y*-coordinate.

$R(2, 1) \rightarrow R'\left(\boxed{}, \boxed{}\right)$

$S(-1, 3) \rightarrow S'\left(\boxed{}, \boxed{}\right)$

$T(-2, 0) \rightarrow T'\left(\boxed{}, \boxed{}\right)$

Check

Quadrilateral *ABCD* has coordinates *A*(−2, 4), *B*(1, 4), *C*(−1, 1), and *D*(−5, 1). The quadrilateral is reflected across the *y*-axis. Write the coordinate notation for a reflection across the *y*-axis. Then, write the coordinates of *A'B'C'D'*.

Part A Write the coordinate notation for a reflection across the *y*-axis.

Part B Write the coordinates of *A'B'C'D'*.

Show your work here

 Talk About It!

Explain why you can multiply each *y*-coordinate by −1 to determine the *y*-coordinate of the image.

 Talk About It!

Explain why *both* the *x*- and *y*-coordinates of *T* and *T'* are the same.

Go Online You can complete an Extra Example online.

Example 4 Describe Reflections

The coordinates of △EFG and its image are shown.

$E(4, 5) \rightarrow E'(4, -5)$

$F(2, 2) \rightarrow F'(2, -2)$

$G(5, 1) \rightarrow G'(5, -1)$

Describe the transformation using words.

Compare the coordinates of the preimage to the coordinates of the image. The _____ -coordinates are the same. The _____ -coordinates are opposites.

The reflection can be described in general notation as

$(x, y) \rightarrow \left(\boxed{} , \boxed{} \right).$

So, the transformation of △EFG is a reflection across the x-axis.

Check

The coordinates of △PQR and its image are shown. Describe the transformation using words.

$P(-3, -2) \rightarrow P'(-3, 2)$

$Q(-4, -3) \rightarrow Q'(-4, 3)$

$R(-2, -4) \rightarrow R'(-2, 4)$

Show your work here

🔾 **Go Online** You can complete an Extra Example online.

📖 **Foldables** It's time to update your Foldable, located in the Module Review, based on what you learned in this lesson. If you haven't already assembled your Foldable, you can find the instructions on page FL1.

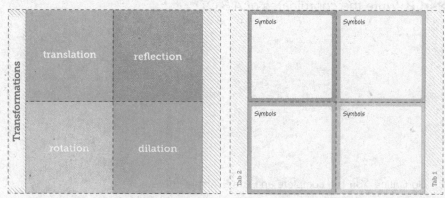

Practice

🔾 **Go Online** You can complete your homework online.

1. The graph of △ABC is shown. Graph the image of △ABC after a reflection across the x-axis. Write the coordinates of the reflected image. (Example 1)

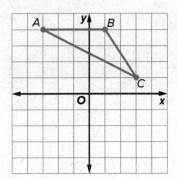

2. The graph of trapezoid WXYZ is shown. Graph the image of WXYZ after a reflection across the y-axis. Write the coordinates of the reflected image. (Example 1)

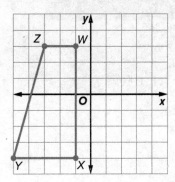

3. The graph of △CDE is shown. Graph the image of △CDE after a reflection across the line x = 2. Include the line of reflection. Then write the coordinates of the image. (Example 2)

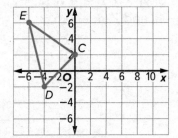

4. The graph of polygon FGHI is shown. Graph the image of FGHI after a reflection across the line y = −1. Include the line of reflection. Then write the coordinates of the image. (Example 2)

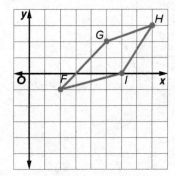

5. Triangle TUV has coordinates T(0, 3), U(−3, 0), and V(−4, 4). The triangle is reflected across the y-axis. Write the coordinate notation for a reflection across the y-axis. Then, write the coordinates of △T′U′V′. (Example 3)

6. The coordinates of △LMN and its image are shown. Describe the transformation. (Example 4)

$$L(0, 0) \rightarrow L'(0, 0)$$

$$M(−4, 1) \rightarrow M'(−4, −1)$$

$$N(−1, 3) \rightarrow N'(−1, −3)$$

7. Equation Editor Triangle XYZ has coordinates $X(-2, 2)$, $Y(-3, -4)$, and $Z(1, -2)$. The triangle is reflected across the x-axis. Write the coordinates of $\triangle X'Y'Z'$.

$X'\left(\boxed{}, \boxed{}\right), Y'\left(\boxed{}, \boxed{}\right), Z'\left(\boxed{}, \boxed{}\right)$

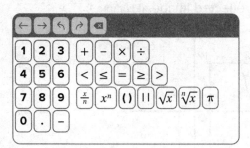

8. MP Identify Structure A polygon and its image after a reflection are shown. Identify the line of reflection. Explain why the images appear to overlap.

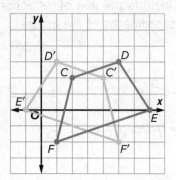

9. MP Find the Error Thomas is finding the coordinates of the image of a polygon with vertices $W(2, 2)$, $X(2, 4)$, $Y(4, 4)$, and $Z(4, 2)$ after a reflection across the y-axis. Describe his error and explain how to correct it.

The coordinates of $W'X'Y'Z'$ are $W'(2, -2)$, $X'(2, -4)$, $Y'(4, -4)$, and $Z'(4, -2)$.

10. MP Persevere with Problems Point T is graphed at $T(-3, 2)$ and is reflected across the line of reflection $y = x$. Write the coordinate notation for a reflection across the line of reflection. Then write the coordinates of T'.

11. Determine whether the following statement is *always*, *sometimes*, or *never* true. Write an argument that can be used to defend your solution.
A preimage and its reflected image are the same shape but different sizes.

Rotations

I Can... use coordinate notation to find the coordinates of a figure that has been rotated about the origin, as well as describe the angle of rotation using the given graph and coordinates of the figures.

What Vocabulary Will You Learn?
center of rotation

rotation

Learn Rotations About a Vertex

A **rotation** is a transformation in which a figure is rotated, or turned, about a fixed point. The **center of rotation** is the fixed point. A rotation does not change the size or shape of the figure. So, the preimage and image are congruent.

Rotations can be described in degrees and direction. The phrases 90° clockwise and 270° counterclockwise are two examples of possible descriptions of rotations.

Go Online Watch the animation to learn about rotating a figure about one of its vertices.

Talk About It!

Describe the image if the line shown is rotated 90° clockwise about the origin.

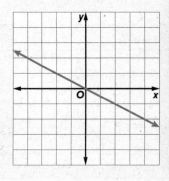

The animation shows two different rotations of △ABC about vertex A.

In a 90° clockwise rotation about vertex A, each point and line segment are rotated that same number of degrees clockwise about the same vertex A.

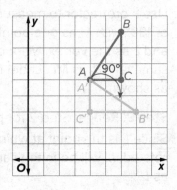

In a 180° counterclockwise rotation about vertex A, each point and line segment are rotated that same number of degrees counterclockwise about the same vertex A.

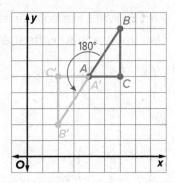

(continued on next page)

The animation also shows the following rotation of △ABC.

In a 270° clockwise rotation about vertex A, each point and line segment are rotated that same number of degrees clockwise about the same vertex A.

Example 1 Rotate Figures About a Vertex

Triangle LMN with vertices L(5, 4), M(5, 7), and N(8, 7) represents a desk in Jackson's bedroom. He wants to rotate the desk counterclockwise 180° about vertex L.

Graph the figure and its image. Then write the coordinates of △L'M'N'.

Part A Graph the figure and its image.

Step 1 Graph the original triangle.

Step 2 Graph the rotated image.

Use a protractor to measure an angle of 180° with M as one point on the ray and L as the vertex. Mark off a point the same length as ML. Label this point M' as shown.

Step 3 Repeat Step 2 for point N. Since L is the point at which △LMN is rotated, L' will be in the same position as L.

Part B Write the coordinates of the image.
Use the graph to write the coordinates of the vertices of the image.

L'(☐ , ☐)

M'(☐ , ☐)

N'(☐ , ☐)

Check

Rectangle *ABCD* with vertices *A*(−7, 4), *B*(−7, 1), *C*(−2, 1), and *D*(−2, 4) represents the bed in Jackson's room. Graph the figure and its image after a clockwise rotation of 90° about vertex *C*. Then write the coordinates of the rectangle *A'B'C'D'*.

Part A Graph the figure and its image after a clockwise rotation of 90° about vertex *C*.

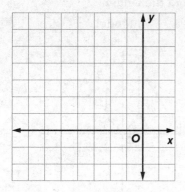

Part B Write the coordinates of rectangle *A'B'C'D'*.

Show your work here

Go Online You can complete an Extra Example online.

Explore Rotate Using Coordinates

Online Activity You will use Web Sketchpad to explore how to rotate figures using coordinates.

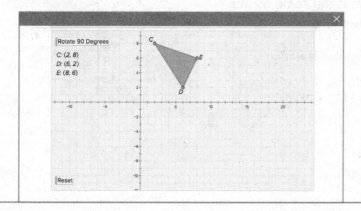

Learn Rotations About the Origin

The coordinates of an image rotated clockwise about the origin can be determined using coordinate notation. When a figure is rotated about the origin, the center of rotation is (0, 0). Each point of the original figure and its image are the same distance from the origin. A rotation's preimage and image are congruent.

90° Clockwise Rotation About the Origin

Words	Model
The x-coordinate of the image is the same as the y-coordinate of the preimage. The y-coordinate of the image is the opposite of the x-coordinate of the preimage.	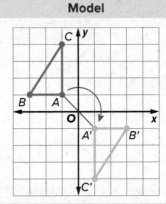

Coordinate Notation

$$(x, y) \rightarrow (y, -x)$$

180° Clockwise Rotation About the Origin

Words	Model
The x- and y-coordinates of the image are opposites of the x- and y-coordinates of the preimage.	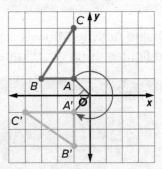

Coordinate Notation

$$(x, y) \rightarrow (-x, -y)$$

270° Clockwise Rotation About the Origin

Words	Model
The x-coordinate of the image is the opposite of the y-coordinate of the preimage. The y-coordinate of the image is the same as the x-coordinate of the preimage.	

Coordinate Notation

$$(x, y) \rightarrow (-y, x)$$

Talk About It!

A classmate claimed that the coordinate notation for the three rotations shown will be the same, even if the center of rotation is not the origin. Do you agree? Justify your response using examples or a counterexample.

Example 2 Rotate Using Coordinates

Triangle *DEF* has vertices *D*(−4, 4), *E*(−1, 2), and *F*(−3, 1). The triangle is rotated 90° clockwise about the origin.

Write the coordinate notation for a clockwise rotation of 90° about the origin. Then write the coordinates of △*D′E′F′*.

Part A Write the coordinate notation for a clockwise rotation of 90° about the origin.

When rotating a point 90° about the origin, the *x*-coordinate of the image is the _____ as the *y*-coordinate of the preimage. The *y*-coordinate of the image is the _____ of the *x*-coordinate of the preimage.

So, the coordinate notation is $(x, y) \rightarrow (y, -x)$.

Part B Write the coordinates of △*D′E′F′*.

$D(-4, 4) \rightarrow D'(\boxed{}, \boxed{})$

$E(-1, 2) \rightarrow E'(\boxed{}, \boxed{})$

$F(-3, 1) \rightarrow F'(\boxed{}, \boxed{})$

Check

Quadrilateral *MNPQ* has coordinates *M*(2, 5), *N*(6, 4), *P*(6, 1) and *Q*(2, 1). The quadrilateral is rotated 180° clockwise about the origin. Write the coordinate notation for a clockwise rotation of 180° about the origin. Then, write the coordinates of quadrilateral *M′N′P′Q′*.

Part A Write the coordinate notation for a clockwise rotation of 180° about the origin.

Part B Write the coordinates of quadrilateral *M′N′P′Q′*.

Show your work here

Go Online You can complete an Extra Example online.

Think About It!

When a figure is rotated 90° clockwise about the origin, how do the *x*-coordinates of the preimage and image compare? the *y*-coordinates?

Talk About It!

The graph shows △*DEF* and its image. How can you verify that this represents a 90° rotation clockwise about the origin?

Example 3 Describe Rotations

Think About It!

How would you begin solving the problem?

Use coordinate notation to describe the rotation. Then determine the angle of rotation. Assume the rotation is clockwise about the origin.

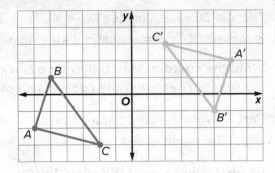

Part A Use coordinate notation to describe the rotation. Compare the coordinates of the preimage and the image.

Preimage	Image
$A\left(\boxed{}, \boxed{}\right)$	$A'\left(\boxed{}, \boxed{}\right)$
$B\left(\boxed{}, \boxed{}\right)$	$B'\left(\boxed{}, \boxed{}\right)$
$C\left(\boxed{}, \boxed{}\right)$	$C'\left(\boxed{}, \boxed{}\right)$

The coordinates of the preimage are the opposites of the coordinates of the image. So, the coordinate notation is $(x, y) \rightarrow (-x, -y)$.

Part B Determine the angle of rotation.

The coordinate notation for a _____ clockwise rotation about the origin is $(x, y) \rightarrow (-x, -y)$. So the angle of rotation is 180°.

Check

Use coordinate notation to describe the rotation. Then determine the angle of rotation. Assume the rotation is clockwise about the origin.

Part A Use coordinate notation to describe the rotation.

Part B Determine the angle of rotation.

Show your work here

Go Online You can complete an Extra Example online.

🌐 **Apply** Arranging Furniture

Before moving furniture in her bedroom, Jasmine made a diagram of the current arrangement. She drew rectangle *ABCD* to represent her desk with vertices *A*(2, 4), *B*(6, 4), *C*(6, 1), and *D*(2, 1). She moved the desk twice, first translating it 3 units left and 2 units down, and then rotating it 90° clockwise about the origin. What are the coordinates of the vertices of the final image after these transformations were applied?

1 What is the task?

Make sure you understand exactly what question to answer or problem to solve. You may want to read the problem three times. Discuss these questions with a partner.

First Time Describe the context of the problem, in your own words.
Second Time What mathematics do you see in the problem?
Third Time What are you wondering about?

2 How can you approach the task? What strategies can you use?

3 What is your solution?

Use your strategy to solve the problem.

4 How can you show your solution is reasonable?

⚓ **Write About It!** Write an argument that can be used to defend your solution.

💬 Talk About It!

Will the results be the same if the figure was rotated first, then translated? Explain.

Check

Before rearranging the furniture in his office, Tyrone made a diagram of the current arrangement. He drew rectangle *ABCD* to represent a file cabinet with vertices *A*(1, 7), *B*(3, 7), *C*(3, 4), and *D*(1, 4). He moved the file cabinet twice, first translating it 4 units right and 3 units down, then rotating it 180° clockwise about the origin. What are the coordinates of the vertices of the final image after these transformations are applied?

Go Online You can complete an Extra Example online.

Foldables It's time to update your Foldable, located in the Module Review, based on what you learned in this lesson. If you haven't already assembled your Foldable, you can find the instructions on page FL1.

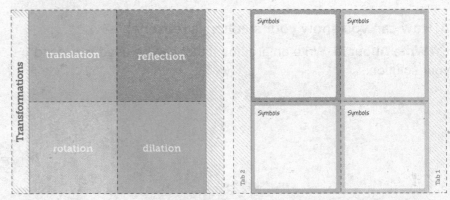

Practice

🔊 **Go Online** You can complete your homework online.

1. Polygon *EFGH* has vertices *E*(−1, 3), *F*(1, 4), *G*(3, 3), and *H*(0, 0). Graph the figure and its image after a clockwise rotation of 90° about vertex *H*. Then write the coordinates of polygon *E'F'G'H'*. (Example 1)

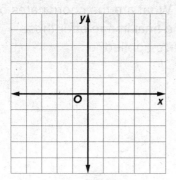

2. Triangle *XYZ* has vertices *X*(−2, −1), *Y*(0, 2), and *Z*(2, −1). Graph the figure and its image after a clockwise rotation of 180° about vertex *Z*. Then write the coordinates of △*X'Y'Z'*. (Example 1)

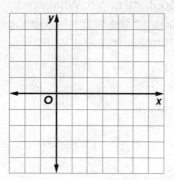

Triangle *QRS* has vertices *Q*(−2, 2), *R*(−3, −4), and *S*(1, −2). Write the coordinate notation for each rotation given. Then write the coordinates of △*Q'R'S'* after each rotation. (Example 2)

3. clockwise rotation of 180° about the origin

4. clockwise rotation of 270° about the origin

5. Use coordinate notation to describe the rotation. Then determine the angle of rotation. Assume the rotation is clockwise about the origin. (Example 3)

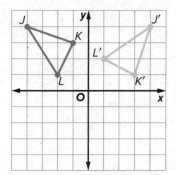

Test Practice

6. **Equation Editor** Point *Z* is located at *Z*(1, −2). Write the coordinates of the point after a clockwise rotation of 90° about the origin.

Z' (⬚ , ⬚)

Apply

7. Before rearranging her dining room furniture, Cathie made a diagram of the current arrangement. She drew rectangle $ABCD$ to represent a china cabinet with vertices $A(2, 8)$, $B(10, 8)$, $C(10, 1)$, and $D(2, 1)$. She moved the cabinet twice, first translating it 2 units right and 1 unit down, then rotating it 90° clockwise about the origin. What are the coordinates of the vertices of the final image after these transformations are applied?

8. Before rearranging his game room, Carlos made a diagram of the current arrangement. He drew rectangle $ABCD$ to represent a tennis table with vertices $A(1, 1)$, $B(1, 12)$, $C(8, 12)$, and $D(8, 1)$. He moved the table twice, first translating it 3 units left and 2 units up, then rotating it 270° clockwise about the origin. What are the coordinates of the vertices of the final image after these transformations are applied?

9. (MP) **Justify Conclusions** A classmate concludes that the image of a figure rotated 270° clockwise will have the same coordinates as the image of the same figure rotated 90° counterclockwise. Is your classmate correct? Write an argument that can be used to defend your solution.

10. (MP) **Model with Mathematics** A figure is rotated 270° counterclockwise about the origin. Then the image is rotated 90° counterclockwise about the origin. Complete the coordinate notation that represents the series of rotations. What can you conclude about the position of the figure after the series of rotations?

$(x, y) \rightarrow \left(\boxed{}, \boxed{} \right) \rightarrow \left(\boxed{}, \boxed{} \right)$

11. (MP) **Reason Inductively** Determine whether the following statement is *always*, *sometimes*, or *never* true. Write an argument that can be used to defend your solution.

    A figure and its rotated image will have the same area, but different perimeters.

Dilations

I Can... describe dilations using coordinate notations as well as graph dilations on the coordinate plane using coordinate notation.

What Vocabulary Will You Learn?

center of dilation

dilation

scale factor

Learn Dilations and Scale Factor

A **dilation** is a transformation which is similar to a scale drawing. It uses a **scale factor** to enlarge or reduce a figure proportionally. Scale factor is the ratio of the side lengths of the image to the side lengths of the preimage.

The preimage and the image are the same shape, but not necessarily the same size. If the scale factor is greater than one, the image is enlarged. If the scale factor is between 0 and 1, the image is reduced. If the scale factor is equal to one, the image is the same size as the preimage.

The scale factor of the dilation shown is found using the ratio of a side length of $\triangle A'B'C'$ to a side length of $\triangle ABC$. The notation CB represents the length of the line segment with endpoints C and B.

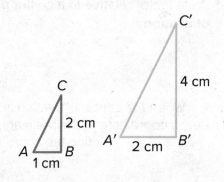

$\dfrac{C'B'}{CB} = \dfrac{4}{2}$ or ☐

The ratio of $C'B'$ to CB is 4:2, or 2. This means the scale factor used to dilate the figure is 2. Since the scale factor is greater than 1, the image was enlarged.

Triangle RST was dilated. The image is triangle $R'S'T'$. To find the scale factor, write the ratio of the image's side length of $R'S'$ to the preimage's side length of RS.

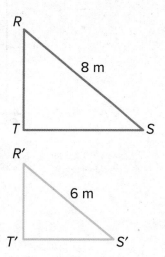

$\dfrac{R'S'}{RS} = \dfrac{6}{8}$ or $\dfrac{3}{4}$

The ratio is 3:4 or $\dfrac{3}{4}$. This means the scale factor used to dilate the figure is between 0 and 1. So, the dilation was a reduction.

💬 **Talk About It!**

Is the ratio of $A'B'$ to AB equivalent to the ratio of $C'B'$ to CB? Explain.

Explore Dilate Figures on the Coordinate Plane

⬆ Online Activity You will use Web Sketchpad to explore how to dilate figures on the coordinate plane when the origin is the center of dilation.

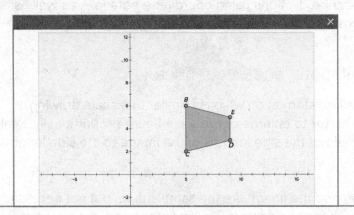

Learn Dilations on a Coordinate Plane

A dilation is a transformation that enlarges or reduces a figure by a scale factor relative to a center point. That point is called the **center of dilation.**

💬 Talk About It!

On the model, why is $k = 2$?

Words
When the center of dilation in the coordinate plane is the origin, each coordinate of the preimage is multiplied by the scale factor k to find the coordinates of the image.

Graph
$(x, y) \rightarrow (2x, 2y)$

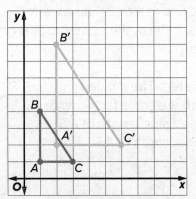

Variables
$(x, y) \rightarrow (kx, ky)$

(continued on next page)

The values for the scale factor *k* determine whether the dilation is an enlargement, a reduction, or if the dilation does not alter the size.

Enlargement	
Words	**Model**
A dilation with a scale factor of *k* will be an image larger than the original if $k > 1$.	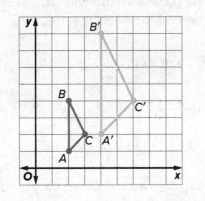
Symbols	
$(x, y) \rightarrow (2x, 2y)$	

Reduction	
Words	**Model**
A dilation with a scale factor of *k* will be an image smaller than the original if $0 < k < 1$.	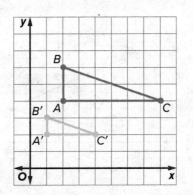
Symbols	
$(x, y) \rightarrow \left(\frac{1}{2}x, \frac{1}{2}y\right)$	

No Change	
Words	**Model**
A dilation with a scale factor of *k* will be an image the same size as the original if $k = 1$.	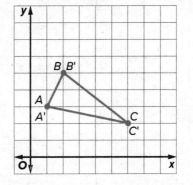
Symbols	
$(x, y) \rightarrow (1x, 1y)$	

💬 **Talk About It!**

Determine whether each scale factor enlarges, reduces, or keeps a figure the same size. Explain.

$\frac{2}{3}$, 6, 1

Think About It!

Is the dilation an enlargement or a reduction?

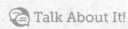

Talk About It!

Compare and contrast the image and the preimage.

Example 1 Graph Dilations

Triangle *ABC* has vertices *A*(−2, 1), *B*(−4, 5), and *C*(3, 2).

Graph the image of the figure after a dilation with a scale factor of 2.

Step 1 Find the coordinates of the image.

The coordinate notation of the dilation is $(x, y) \rightarrow (2x, 2y)$. Multiply the coordinates of each vertex by 2.

$A(-2, 1) \rightarrow (2 \cdot -2, 2 \cdot 1) \rightarrow A'\left(\boxed{}, \boxed{}\right)$

$B(-4, 5) \rightarrow (2 \cdot -4, 2 \cdot 5) \rightarrow B'\left(\boxed{}, \boxed{}\right)$

$C(3, 2) \rightarrow (2 \cdot 3, 2 \cdot 2) \rightarrow C'\left(\boxed{}, \boxed{}\right)$

Step 2 Use the coordinates of *A'B'C'* to graph the image on the coordinate plane.

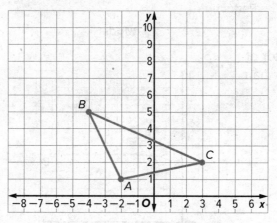

Check

Triangle *DEF* has vertices *D*(−2, −1), *E*(0, 1), and *F*(1, −3). Graph the triangle and its image after a dilation with a scale factor of 3.

Show your work here

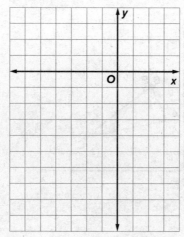

Go Online You can complete an Extra Example online.

Example 2 Graph Dilations

Triangle *JKL* has vertices *J*(3, 8), *K*(10, 6), and *L*(8, 2).

Graph the image after a dilation with a scale factor of $\frac{1}{2}$.

Step 1 Find the coordinates of the image.

The coordinate notation of the dilation is $(x, y) \rightarrow \left(\frac{1}{2}x, \frac{1}{2}y\right)$.
Multiply the coordinates of each vertex by $\frac{1}{2}$.

$$J(3, 8) \rightarrow \left(\frac{1}{2} \cdot 3, \frac{1}{2} \cdot 8\right) \rightarrow J'\left(\frac{3}{2}, 4\right)$$

$$K(10, 6) \rightarrow \left(\frac{1}{2} \cdot 10, \frac{1}{2} \cdot 6\right) \rightarrow K'\left(\boxed{}, \boxed{}\right)$$

$$L(8, 2) \rightarrow \left(\frac{1}{2} \cdot 8, \frac{1}{2} \cdot 2\right) \rightarrow L'\left(\boxed{}, \boxed{}\right)$$

Step 2 Use the coordinates of *J'K'L'* to graph the image on the coordinate plane.

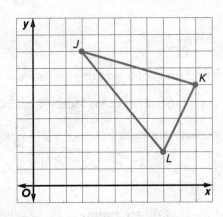

Check

Triangle *EFG* has vertices *E*(−6, 9), *F*(3, 6), and *G*(−3, 3). Graph the triangle and its image after a dilation with a scale factor of $\frac{1}{3}$.

Show your work here

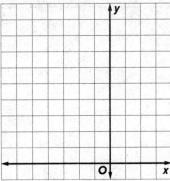

🅝 **Go Online** You can complete an Extra Example online.

Think About It!

Is the dilation an enlargement or a reduction?

Talk About It!

Compare and contrast the image and the preimage.

Lesson 8-4 · Dilations **469**

Think About It!

Which figure is the preimage? How does the image compare in size to the preimage?

Example 3 Describe Dilations

Use coordinate notation to describe the dilation.

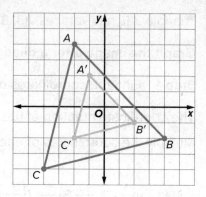

Compare the coordinates of the preimage and the image.

$A\left(\boxed{}, \boxed{}\right) \rightarrow A'\left(\boxed{}, \boxed{}\right)$

$B\left(\boxed{}, \boxed{}\right) \rightarrow B'\left(\boxed{}, \boxed{}\right)$

$C\left(\boxed{}, \boxed{}\right) \rightarrow C'\left(\boxed{}, \boxed{}\right)$

The coordinates of the image are half of the coordinates of the preimage. The scale factor is $\frac{1}{2}$. So, the dilation is a reduction and the coordinate notation is $(x, y) \rightarrow \left(\frac{1}{2}x, \frac{1}{2}y\right)$.

Check

Use coordinate notation to describe the dilation.

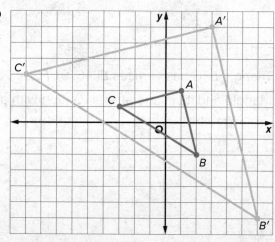

🅝 **Go Online** You can complete an Extra Example online.

🌐 Apply Consumer Science

The graph shows the plans for a new fence that Olivia is building on her horse farm. Each unit on the graph represents 8 feet of fencing. After studying the plans, Olivia decides she would like to build a fence that encloses a greater area. If Olivia dilates Rectangle *ABCD* by a scale factor of 2, and fencing costs $12.50 per foot, how much will she spend on fencing?

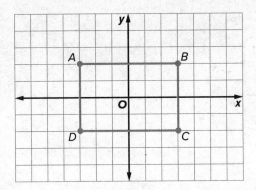

🔵 **Go Online**
Watch the animation.

1 What is the task?

Make sure you understand exactly what question to answer or problem to solve. You may want to read the problem three times. Discuss these questions with a partner.

First Time Describe the context of the problem, in your own words.
Second Time What mathematics do you see in the problem?
Third Time What are you wondering about?

2 How can you approach the task? What strategies can you use?

3 What is your solution?

Use your strategy to solve the problem.

4 How can you show your solution is reasonable?

🔺 **Write About It!** Write an argument that can be used to defend your solution.

💬 **Talk About It!**

Compare the perimeters of the preimage and the image. What do you notice? How does this relate to scale factor?

Check

The graph shows the plans for a new fence that Oliver is building in his back yard for his dog. Each unit on the graph represents 10 feet of fencing. After studying the plans, he decides he would like to build a fence that encloses a smaller area. If Oliver dilates Rectangle *ABCD* by a scale factor of 0.75, and fencing costs $8.30 per foot, how much will he spend on fencing?

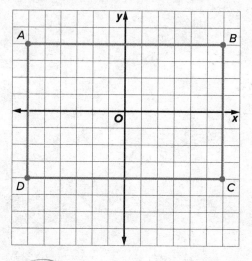

Show your work here

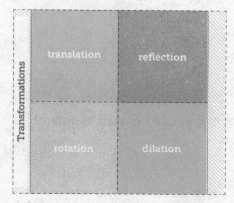Go Online You can complete an Extra Example online.

 Foldables It's time to update your Foldable, located in the Module Review, based on what you learned in this lesson. If you haven't already assembled your Foldable, you can find the instructions on page FL1.

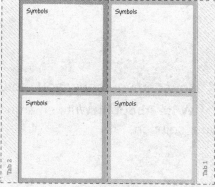

Practice

Go Online You can complete your homework online.

1. Trapezoid *RAIN* has vertices *R*(−2, 1), *A*(1, 1), *I*(0, −1), and *N*(−1, −1). Graph the image of the figure after a dilation with a scale factor of 2. (Example 1)

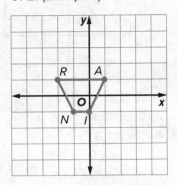

2. Triangle *ABC* has vertices *A*(2, 1), *B*(3, 0), and *C*(1, −2). Graph the image of the figure after a dilation with a scale factor of 3. (Example 1)

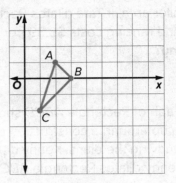

3. Triangle *JKL* has vertices *J*(−4, −1), *K*(0, 4), and *L*(−4, −2). Graph the image of the figure after a dilation with a scale factor of 0.5. (Example 2)

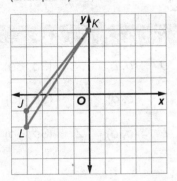

4. Rectangle *PQRS* has vertices *P*(−3, 3), *Q*(6, 3), *R*(6, −3), and *S*(−3, −3). Graph the image of the figure after a dilation with a scale factor of $\frac{1}{3}$. (Example 2)

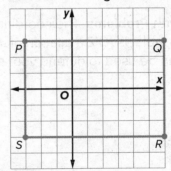

Test Practice

5. Use coordinate notation to describe the dilation. (Example 3)

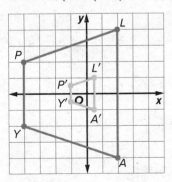

6. **Equation Editor** Keisha used a photo that measured 4 inches by 6 inches to make a copy that measured 8 inches by 12 inches. What is the scale factor of the dilation?

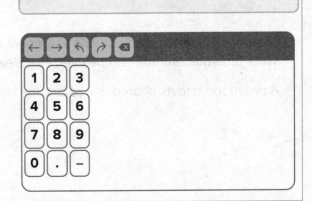

Apply

For Exercises 7 and 8, use the graph of Rectangle *ABCD*.

7. Suppose the graph represents the plans for a fence that Tara is building for a new city dog park. Each unit on the graph represents 12 yards. After studying the plans, Tara decides to build a fence that encloses a smaller area. If Tara dilates Rectangle *ABCD* by a scale factor of 0.75, and fencing costs $6.39 per yard, how much will she spend on fencing?

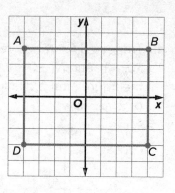

8. Suppose the graph of Rectangle *ABCD* shows the plans for a safety fence that Kenny is setting up around a construction area. Each unit on the graph represents 25 feet. After studying the plans, Kenny decides to build a fence that encloses a larger area. If Kenny dilates Rectangle *ABCD* by a scale factor of 2.5, and fencing costs $5.25 per foot, how much will he spend on fencing?

9. Ⓜ️ **Persevere with Problems** The coordinates of two triangles are shown in the table. Is *XYZ* a dilation of *JKL*? Write an argument that can be used to defend your solution.

△*JKL*		△*XYZ*	
J	(*a*, *b*)	*X*	(3*a*, 6*b*)
K	(*c*, *d*)	*Y*	(3*c*, 6*d*)
L	(*a*, *d*)	*Z*	(3*a*, 6*d*)

10. Ⓜ️ **Find the Error** Kelly is finding the coordinates of the image of a polygon with vertices *W*(2, 2), *X*(2, 4), *Y*(4, 4), and *Z*(4, 2) after a dilation with a scale factor of 3. Describe her error and explain how to correct it.

The coordinates of *W'X'Y'Z'* are *W'*(2, 6), *X'*(2, 12), *Y'*(4, 12), and *Z'*(4, 6).

11. Determine whether the following statement is *always*, *sometimes*, or *never* true. Write an argument that can be used to defend your solution.

A preimage and its dilated image are the same shape but different sizes.

📖 **Foldables** Use your Foldable to help review the module.

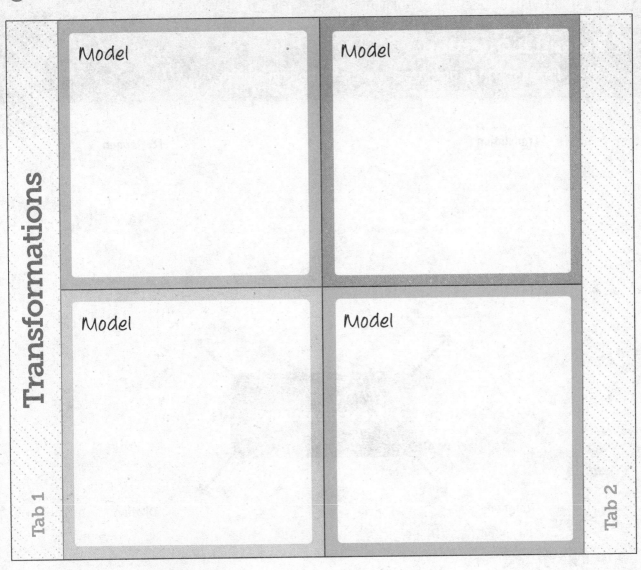

Transformations

Tab 1

Model

Model

Model

Model

Tab 2

Rate Yourself! ⬛ ◆ ★

Complete the chart at the beginning of the module by placing a checkmark in each row that corresponds with how much you know about each topic after completing this module.

Write about one thing you learned.

_____

_____

_____

Write about a question you still have.

_____

_____

_____

Reflect on the Module

Use what you learned about transformations to complete the graphic organizer.

e Essential Question

What does it mean to perform a transformation on a figure?

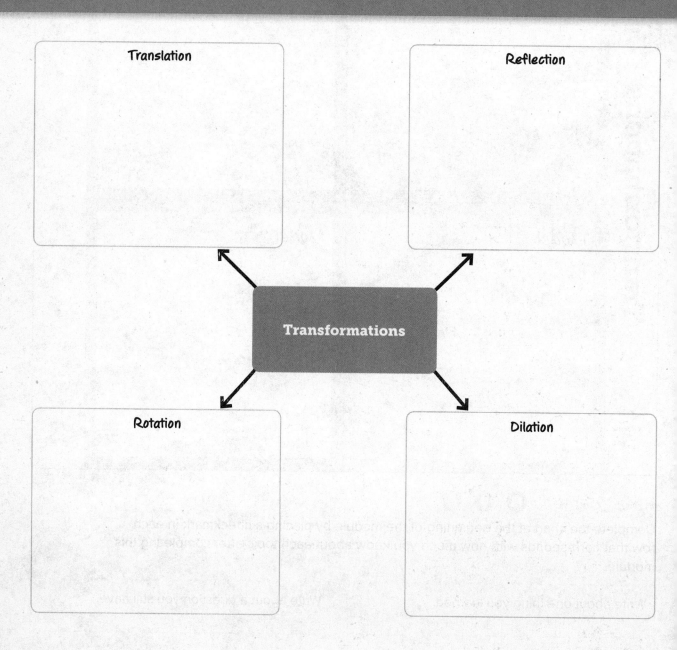

Translation

Reflection

Transformations

Rotation

Dilation

Test Practice

1. Grid The graph of △ABC is shown. (Lesson 1)

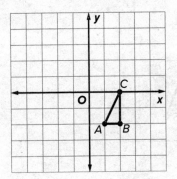

A. Graph the image of △ABC after a translation of 3 units left and 4 units up.

B. Write the coordinates of the image, △A'B'C'.

_____

2. Multiple Choice Which of the following coordinate notations describes the translation shown? (Lesson 1)

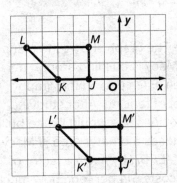

Ⓐ $(x, y) \rightarrow (x + 6, y - 7)$

Ⓑ $(x, y) \rightarrow (x + 2, y - 7)$

Ⓒ $(x, y) \rightarrow (x + 2, y - 5)$

Ⓓ $(x, y) \rightarrow (x - 2, y + 5)$

3. Multiselect Consider the graph of △ABC and its image, △A'B'C'. Which of the following statements is accurate regarding the transformation shown? Select all that apply. (Lesson 2)

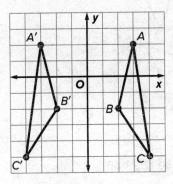

☐ △A'B'C' is the image of a reflection of △ABC across the x-axis.

☐ △A'B'C' is the image of a reflection of △ABC across the y-axis.

☐ The x-coordinates of the vertices of △ABC and its image △A'B'C' are the same.

☐ The x-coordinates of the vertices of △ABC and its image △A'B'C' are opposites.

☐ The y-coordinates of the vertices of △ABC and its image △A'B'C' are the same.

☐ The y-coordinates of the vertices of △ABC and its image △A'B'C' are opposites.

4. Open Response Triangle DEF has coordinates D(2, 3), E(6, 1), and F(2, 0). The triangle is reflected across the x-axis. Write the coordinates of △D'E'F'. (Lesson 2)

_____

5. Multiple Choice Which of the following statements describes the rotation shown? Assume the rotation is clockwise about the origin. (Lesson 3)

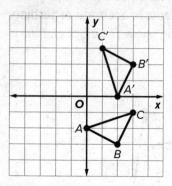

Ⓐ 90° clockwise rotation

Ⓑ 180° clockwise rotation

Ⓒ 270° clockwise rotation

Ⓓ 360° clockwise rotation

6. Grid Triangle *JKL* with vertices *J*(4, 4), *K*(4, 6), and *L*(1, 6) represents an end table in Stacey's family room. She wants to rotate the end table counterclockwise 180° about vertex *J*. (Lesson 3)

A. Graph the figure and its image on the coordinate plane.

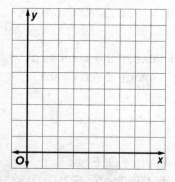

B. Write the coordinates of △*J'K'L'*.

7. Multiselect Which of the following statements accurately describe the dilation shown? Select all that apply. (Lesson 4)

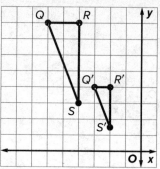

☐ The coordinates of △*Q'R'S'* are half that of the coordinates of △*QRS*.

☐ The coordinates of △*Q'R'S'* are twice the coordinates of △*QRS*.

☐ The dilation is an enlargement.

☐ The dilation is a reduction.

☐ The coordinate notation for the dilation is $(x, y) \rightarrow \left(\frac{1}{2}x, \frac{1}{2}y\right)$.

☐ The coordinate notation for the dilation is $(x, y) \rightarrow (2x, 2y)$.

8. Open Response The graph shows the plans for a fence that Macy is building around her vegetable garden. Each unit on the graph represents 5 feet of fencing. After studying her plans, she decides she would like to build a fence that encloses a greater area. If Macy dilates Rectangle *EFGH* by a scale factor of 1.5, and fencing costs $14 per foot, how much will she spend on fencing? (Lesson 4)

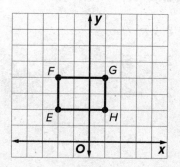

Module 9

Congruence and Similarity

e Essential Question

What information is needed to determine if two figures are congruent or similar?

What Will You Learn?

Place a checkmark (✓) in each row that corresponds with how much you already know about each topic **before** starting this module.

	Before			After		
KEY — ◼ I don't know. ◓ I've heard of it. ★ I know it!	◼	◓	★	◼	◓	★
determining whether figures are congruent						
identifying which sequence of transformations maps one figure onto a congruent figure						
writing congruence statements						
using properties of congruent figures to determine missing measures in figures						
determining whether figures are similar						
identifying which sequence of transformations maps one figure onto a similar figure						
using scale factors						
writing similarity statements						
showing that two triangles are similar using Angle-Angle Similarity						
using properties of similar figures to determine missing measures in figures						
solving problems using indirect measurement						

📭 **Foldables** Cut out the Foldable and tape it to the Module Review at the end of the module. You can use the Foldable throughout the module as you learn about congruence and similarity.

What Vocabulary Will You Learn?

Check the box next to each vocabulary term that you may already know.

☐ Angle-Angle Similarity

☐ composition of transformations

☐ congruent

☐ corresponding parts

☐ indirect measurement

☐ similar

Are You Ready?

Study the Quick Review to see if you are ready to start this module.
Then complete the Quick Check.

Quick Review

Example 1

Find equivalent ratios.

Find the value of m in the equation $\frac{m}{8} = \frac{3}{4}$.

$\frac{m}{8} = \frac{3}{4}$ Find an equivalent ratio.

$\frac{6}{8} = \frac{3}{4}$ Muliply 3 by 2.

So, $m = 6$.

Example 2

Transform figures.

Triangle ABC has vertices $A(1, 0)$, $B(3, 4)$, and $C(6, 2)$. What are the coordinates of vertex B after the triangle is translated 4 units left and 3 units down?

$B(3, 4)$ Write the coordinate of point B.

$B'(3 - 4, 4 - 3)$ Subtract 4 from the x-coordinate and subtract 3 from the y-coordinate.

$B'(-1, 1)$ Simplify.

Quick Check

1. Sallie earned $130 in 4 weeks. Solve $\frac{130}{4} = \frac{x}{10}$ to find how much, in dollars x, Sallie would earn in 10 weeks at this rate.

2. Triangle XYZ has vertices $X(-2, 1)$, $Y(2, 3)$, and $Z(0, -5)$. What are the coordinates of vertex X after the triangle is reflected across the y-axis?

How Did You Do?

Which exercises did you answer correctly in the Quick Check?
Shade those exercise numbers at the right.

Congruence and Transformations

I Can... use a composition of transformations, as well as the orientations of figures, to determine if two figures are congruent.

What Vocabulary Will You Learn?
composition of transformations

congruent

Explore Congruence and Transformations

Online Activity You will use Web Sketchpad to explore properties of translations, reflections, and rotations.

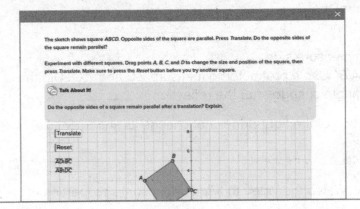

Learn Congruence and Transformations

Translations, reflections, and rotations preserve the shape and size of a figure.

Go Online Watch the video to learn about some properties of translations.

The video shows animation frames using a stack of index cards. A parallelogram is traced on each card.

As you flip through the cards from front to back, the parallelogram appears to move. As the parallelogram moves, the sides of the parallelogram remain the same length. The measures of the angles also remain the same. So, sliding or moving a figure does not change its shape or size.

(continued on next page)

Go Online Watch the video to learn about some properties of reflections.

The video shows how to reflect a triangle using tracing paper.

Step 1 Draw △ABC on tracing paper. Draw a dotted line (the line of reflection) on the paper as shown.

Step 2 Fold the paper along the dotted line. Trace the triangle onto the folded portion of the tracing paper. Unfold and label the vertices A′, B′, and C′.

Use a ruler to measure side AB and side A′B′. Use a protractor to measure ∠C and ∠C′. Did the size of the triangle change after the reflection? _____

So, the reflection of the figure does not change its shape or size.

Go Online Watch the video to learn about some properties of rotations.

The video shows how to rotate a parallelogram using tracing paper.

Step 1 Place a piece of tracing paper over parallelogram WXYZ shown. Copy the parallelogram. Trace points A, B, C, and $\overrightarrow{AB}$.

Step 2 Place the eraser end of your pencil on point A. Turn the tracing paper to the right until $\overrightarrow{AB}$ passes through point C.

Use a ruler to measure side WX and its corresponding side on the image. Use a protractor to measure ∠Y and its corresponding angle on the image. Did the size of the parallelogram change after the rotation? _____

So, the rotation of the figure does not change its shape or size.

(continued on next page)

When a transformation is applied to a figure and then another transformation is applied to the image, the result is a **composition of transformations**, or a sequence of transformations.

On the graph, triangle *ABC* is reflected across the *y*-axis to create triangle *A'B'C'*. Then triangle *A'B'C'* is rotated 90° clockwise about the origin to create triangle *A"B"C"*. The symbol ″ is read *double prime* and is used to indicate a second transformation of a figure.

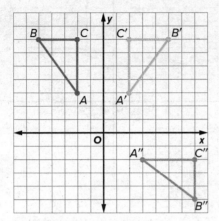

You can show two figures are **congruent** if the second can be obtained from the first by a sequence of rotations, reflections, and/or translations.

Since a sequence of translations, reflections, and/or rotations does not change the shape or size of a figure, the image and preimage are congruent. So, line segments in the preimage have the same length as line segments in the image. Angles in the preimage have the same measure as angles in the image.

(continued on next page)

Pause and Reflect

How are the terms *congruent* and *composition of transformations* related?

Record your observations here

Go Online Watch the animation to see how two figures are congruent.

The animation shows that you can use transformations to determine if △DEF is congruent to △HGI.

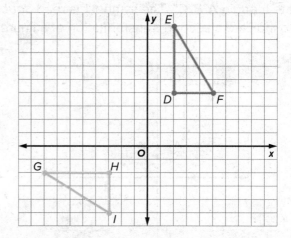

Step 1 Rotate △DEF 90° clockwise about vertex D.

Step 2 Reflect △D'E'F' across the y-axis.

Step 3 Translate △D"E"F" to the left one unit and down six units so that it maps exactly onto △HGI.

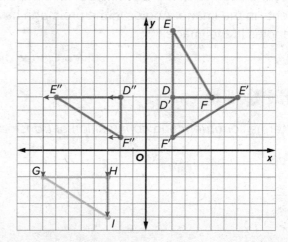

So, the triangles are congruent because △DEF can be mapped onto △HGI by using a sequence of a rotation, a reflection, and a translation.

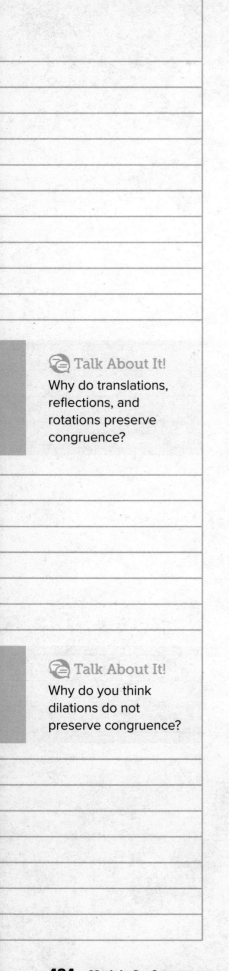

Talk About It!

Why do translations, reflections, and rotations preserve congruence?

Talk About It!

Why do you think dilations do not preserve congruence?

Example 1 Determine Congruence

Are the two figures congruent? If so, describe a sequence of transformations that maps △ABC onto △XYZ. If not, explain why they are not congruent.

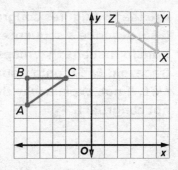

Part A Determine if the two figures are congruent by using transformations.

Reflect △ABC across the y-axis. Then translate △A'B'C' up 4 units. Triangle ABC is mapped onto △XYZ.

Since △ABC is mapped onto △XYZ with a reflection followed by a translation, the two triangles are congruent.

Part B Describe the sequence of transformations.

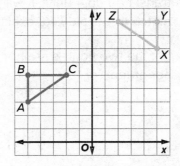

A reflection across the _____-axis followed by a translation _____ units up maps △ABC onto △XYZ.

💬 Talk About It!

Can you translate △ABC up 4 units first and then reflect it across the y-axis? Explain. Why is △ABC congruent to △XYZ?

Check

Refer to Figure A and Figure B.

Part A Determine if the figures are congruent.

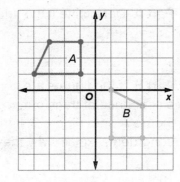

Part B If the figures are congruent, describe a sequence of transformations that maps Figure A onto Figure B. If the figures are not congruent, explain why they are not congruent.

Show your work here

🅝 **Go Online** You can complete an Extra Example online.

Example 2 Determine Congruence

Think About It!

Do the trapezoids appear congruent? Why or why not?

Are the two figures congruent? If so, describe a sequence of transformations that maps trapezoid _EFGH_ onto trapezoid _IJKL_. If not, explain why they are not congruent.

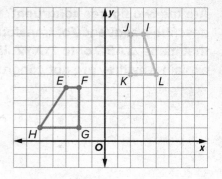

Reflect trapezoid _EFGH_ across the _y_-axis. Even if the reflected figure is translated up 4 units, it will not match trapezoid _IJKL_ exactly.

So, the two figures are not congruent.

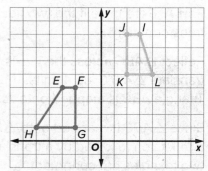

Check

Are the two figures congruent? If so, describe a sequence of transformations that maps Figure A onto Figure B. If not, explain why they are not congruent.

Part A Determine if the figures are congruent.

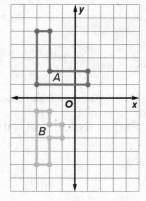

Part B If the figures are congruent, describe a sequence of transformations that maps Figure A onto Figure B. If the figures are not congruent, explain why they are not congruent.

Show your work here

🐢 **Go Online** You can complete an Extra Example online.

Learn Identify Transformations

The order in which the vertices of a figure are named determines the figure's orientation. In the reflection shown, the vertices of the preimage are named in a clockwise direction, but the vertices of the image are named in a counterclockwise direction. The orientation has been reversed.

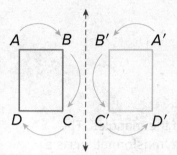

If you have two congruent figures, you can determine the transformation, or sequence of transformations, that maps one figure onto the other by analyzing the orientation of the figures.

Translation	Reflection
• length is the same • orientation is the same	• length is the same • orientation is reversed

Rotation
• length is the same • orientation is the same

Example 3 Identify Transformations

Triangle *ABC* is congruent to △*XYZ*. Determine which sequence of transformations maps △*ABC* onto △*XYZ*.

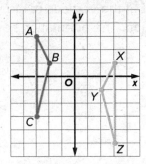

Determine any changes in the orientation of the triangles. The orientation is reversed so at least one of the transformations is a reflection.

If you reflect △*ABC* across the *y*-axis and then translate it down 2 units, it coincides with △*XYZ*.

So, the transformations that map △*ABC* onto △*XYZ* consist of a

_____ across the *y*-axis followed by a _____

2 units down.

Check

Parallelogram *KLMN* is congruent to parallelogram *WXYZ*. Which sequence of transformations maps parallelogram *KLMN* onto parallelogram *WXYZ*?

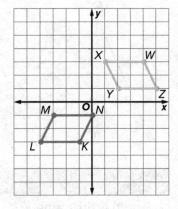

Ⓐ Translate parallelogram *KLMN* 4 units up and then translate it 4 units to the left.

Ⓑ Reflect parallelogram *KLMN* across the *y*-axis and then reflect it across the *x*-axis.

Ⓒ Rotate parallelogram *KLMN* 180° counterclockwise about the origin and then translate it 2 units to the right.

Ⓓ Reflect parallelogram *KLMN* across the *x*-axis and then translate it 5 units to the right.

Show your work here

Think About It!

Compare and contrast △*ABC* and △*XYZ*. What do you notice?

Talk About It!

How does analyzing the orientation of two figures help determine what transformations map one figure onto another?

 Go Online You can complete an Extra Example online.

🌐 Example 4 Identify Transformations

Ms. Martinez created the logo shown for Diamond Plumbing.

What transformations did she use if the letter "d" is the preimage and the letter "p" is the image? Are the two figures congruent?

Part A Identify the transformations.

Step 1 Start with the preimage.
Determine which transformation can be used to create the letter "p".

Step 2 Rotate the letter "d" 180° about point *A*.

Step 3 Translate the new image up.

So, Ms. Martinez could have used a rotation and a translation to create the logo.

Part B Determine congruence.

The letters are congruent because images produced by a

_____ and a _____ have the same shape and size.

Check the solution.

Copy the logo onto a piece of paper. Then trace the letter "d" with tracing paper. Rotate the letter 180° about Point *A*. Slide it up to line up with the letter "p". The letters are the same shape and size. They are congruent.

💭 **Think About It!**

Compare and contrast the letters "d" and "p" in the logo. What do you notice?

💬 **Talk About It!**

Describe another set of transformations that can map the letter "d" onto the letter "p".

Check

Matthew designed the logo shown.

Part A What transformations could be used to create the logo if the letter "W" is the preimage and the letter "M" is the image?

Ⓐ a translation

Ⓑ a reflection followed by a dilation

Ⓒ a rotation followed by a dilation

Ⓓ a reflection followed by a translation

Part B Are the two figures congruent?

Show your work here

🡒 **Go Online** You can complete an Extra Example online.

📙 **Foldables** It's time to update your Foldable, located in the Module Review, based on what you learned in this lesson. If you haven't already assembled your Foldable, you can find the instructions on page FL1.

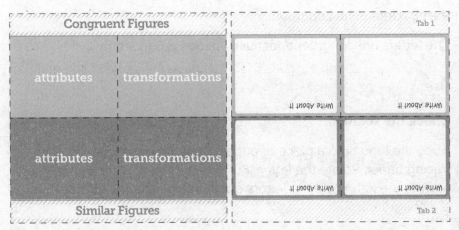

Practice

🡒 **Go Online** You can complete your homework online.

Determine if each pair of figures are congruent. If so, describe a sequence of transformations that maps one figure onto the other figure. If not, explain why they are not congruent. (Examples 1 and 2)

1.

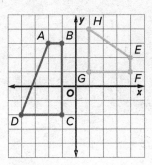

2.

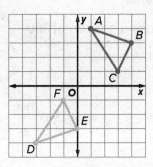

3. Parallelogram *CAMP* is congruent to parallelogram *SITE*. Determine which sequence of transformations maps parallelogram *CAMP* onto parallelogram *SITE*. (Example 3)

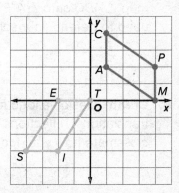

4. For his school web page, Manuel created the logo shown at the right. What transformations could be used to create the logo if Figure A is the preimage and Figure B is the image? Are the two figures congruent? (Example 4)

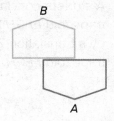

5. For the local art gallery opening, the curator had the design shown at the right created. What transformations could be used to create the design if Figure A is the preimage and Figure B is the image? Are the two figures congruent? (Example 4)

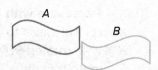

6. **Multiple Choice** Trapezoid *QRST* and its image are shown. What transformation maps trapezoid *QRST* onto trapezoid *LMNO*?

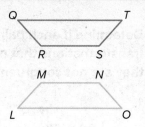

Ⓐ dilation about vertex *R*

Ⓑ vertical translation

Ⓒ reflection across a horizontal line

Ⓓ rotation about vertex *Q*

Apply

7. Ⓜ **Identify Structure** In some cases, a sequence of transformations is the same as a single transformation. Triangle *ABC* is reflected across the *y*-axis, and then reflected across the *x*-axis. Is there a single transformation that would map △*ABC* onto △*A"B"C"*? Write an argument that can be used to defend your solution.

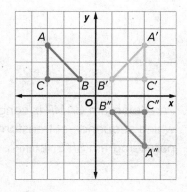

8. **Create** Design a logo for a club at your school, using translations, reflections, and/or rotations. Then explain to a classmate how your logo uses congruent figures.

9. A student concluded that trapezoid *ABCD* is congruent to trapezoid *EFGH* because a reflection across the *y*-axis followed by a translation 2 units down maps trapezoid *ABCD* onto trapezoid *EFGH*. Find the student's mistake and correct it.

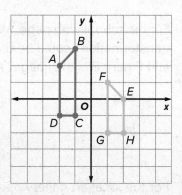

10. Ⓜ **Persevere with Problems** Triangle *XYZ* is reflected across the *x*-axis to produce triangle *X'Y'Z'*. Then triangle *X'Y'Z'* is rotated 90° counterclockwise about the origin to create triangle *X"Y"Z"*. If triangle *X"Y"Z"* has vertices *X"*(4, 0), *Y"*(2, −1), *Z"*(2, 1), what are the coordinates of the vertices of triangle *XYZ*?

Congruence and Corresponding Parts

I Can... use the properties of rotations, reflections, and translations to identify congruent parts of congruent figures and to find missing measures.

What Vocabulary Will You Learn?
corresponding parts

Learn Corresponding Parts of Congruent Figures

In the figure, the two triangles are congruent because △DEF is the image of △ABC reflected across line *m*.

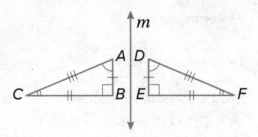

The parts of congruent figures that *match*, or correspond, are called **corresponding parts**.

To indicate that sides are congruent, an equal number of tick marks is drawn on the corresponding sides. To show that angles are congruent, an equal number of arcs is drawn on the corresponding angles. The notation △ABC ≅ △DEF is read *triangle ABC is congruent to triangle DEF.*

Words
If two figures are congruent, their corresponding sides are congruent and their corresponding angles are congruent.

Model
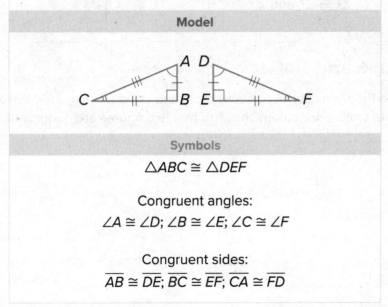

Symbols
△ABC ≅ △DEF
Congruent angles:
∠A ≅ ∠D; ∠B ≅ ∠E; ∠C ≅ ∠F
Congruent sides:
$\overline{AB} \cong \overline{DE}$; $\overline{BC} \cong \overline{EF}$; $\overline{CA} \cong \overline{FD}$

Math History Minute

Third century Chinese mathematician **Liu Hui** proved many of the algorithms that were stated in the famous Chinese text *The Nine Chapters on the Mathematical Art*. He also is known for producing the most accurate estimate of the value of π known to exist in the ancient world. By inscribing a polygon of 192 sides in a circle, he obtained the value 3.141024.

💭 Think About It!

How do the tick marks help you know which parts of each triangle are congruent?

💬 Talk About It!

When writing congruence statements, why is it important to match up corresponding vertices in the statement?

Example 1 Write Congruence Statements

Write congruence statements for the corresponding parts in the congruent triangles shown.

Use the matching arcs and tick marks to identify the corresponding parts.

Corresponding sides:

$\overline{JK}$ corresponds to [ ]

$\overline{KL}$ corresponds to [ ]

$\overline{LJ}$ corresponds to [ ]

So, the congruence statements for the corresponding sides are $\overline{JK} \cong \overline{GH}$, $\overline{KL} \cong \overline{HI}$, and $\overline{LJ} \cong \overline{IG}$.

Corresponding angles:

$\angle J$ corresponds to $\angle$ [ ]

$\angle L$ corresponds to $\angle$ [ ]

$\angle K$ corresponds to $\angle$ [ ]

So, the congruence statements for the corresponding angles are $\angle J \cong \angle G$, $\angle L \cong \angle I$, and $\angle K \cong \angle H$.

Pause and Reflect

Using the terms *corresponding parts* and *congruent figures*, explain how to write a statement showing that two figures are congruent.

Record your observations here

Check

The trapezoids shown are congruent.

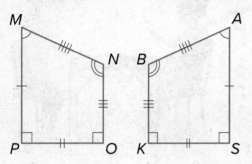

Complete each congruence statement to compare the corresponding parts.

$\angle A \cong$ [   ] $\overline{AB} \cong$ [   ]

$\angle B \cong$ [   ] $\overline{BK} \cong$ [   ]

$\angle K \cong$ [   ] $\overline{KS} \cong$ [   ]

$\angle S \cong$ [   ] $\overline{SA} \cong$ [   ]

Show your work here

🐦 **Go Online** You can complete an Extra Example online.

Pause and Reflect

Did you make any errors when completing the Check exercise? What can you do to make sure you don't repeat that error in the future?

Record your observations here

🌐 Example 2 Find Missing Measures

Liliana is using a brace to support a tabletop. In the figure, △BCE ≅ △DFG.

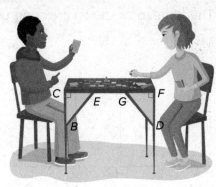

If m∠CEB = 50°, what is the measure of ∠FGD?

You can use the properties of congruent figures to determine missing measures of angles and lengths of sides in a figure.

Step 1 Identify the corresponding parts.

∠CEB and ∠ _____ are corresponding parts in the congruent triangles.

Step 2 Find the missing angle measure.

m∠CEB = m∠FGD   Corresponding angles have the same measure.

$$\boxed{} = m∠FGD$$

So, ∠FGD measures 50°.

Check

A set of windows contains two congruent trapezoids, *ABCD* and *WXYZ*. If m∠B = 110°, what is the measure of ∠X?

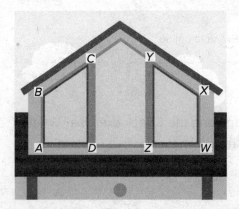

Show your work here

🔵 **Go Online** You can complete an Extra Example online.

🌐 Apply Construction

A diagram of a truss bridge is shown. In the diagram, $\triangle ABC \cong \triangle DEC$. If $AB = 29$ feet and $AC = 33$ feet, what is the length of $\overline{EC}$? Round to the nearest tenth.

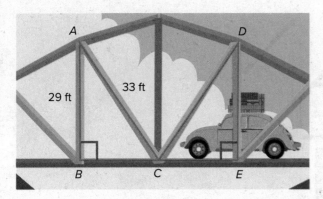

1 What is the task?

Make sure you understand exactly what question to answer or problem to solve. You may want to read the problem three times. Discuss these questions with a partner.

First Time Describe the context of the problem, in your own words.
Second Time What mathematics do you see in the problem?
Third Time What are you wondering about?

2 How can you approach the task? What strategies can you use?

Record your observations here

3 What is your solution?

Use your strategy to solve the problem.

Show your work here

4 How can you show your solution is reasonable?

⚡ **Write About It!** Write an argument that can be used to defend your solution.

💬 **Talk About It!**

How do you know the answer cannot be greater than 33 feet?

Check

A diagram of an attic truss is shown. In the diagram, △ABC ≅ △XYZ. If AB = 5 feet and AC = 6.5 feet, what is the length of $\overline{YZ}$? Round to the nearest tenth.

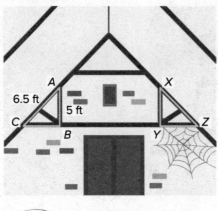

Show your work here

Go Online You can complete an Extra Example online.

Foldables It's time to update your Foldable, located in the Module Review, based on what you learned in this lesson. If you haven't already assembled your Foldable, you can find the instructions on page FL1.

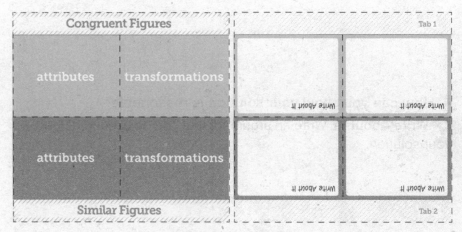

Practice

🡒 Go Online You can complete your homework online.

Write congruence statements comparing the corresponding parts in each set of congruent figures. (Example 1)

1.
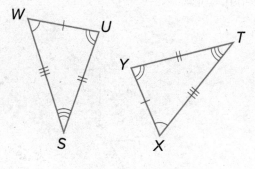

2.
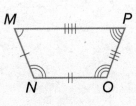

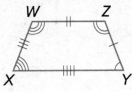

3.
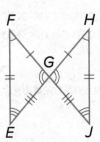

4.
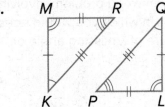

Test Practice

5. In the quilt design shown, △RST ≅ △RWX. If m∠WXR = 62°, what is the measure of ∠STR? (Example 2)

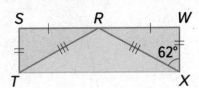

6. **Open Response** In the baseball diamond shown, △BEA ≅ △ARB. The length of $\overline{BE}$ is 90 feet. What is the length of $\overline{AR}$? (Example 2)

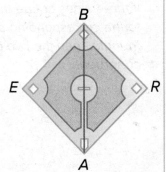

Apply

7. In the roof construction shown, △ABC ≅ △DEF. If AB = 8.5 feet and AC = 10 feet, what is the length of $\overline{EF}$? Round to the nearest tenth.

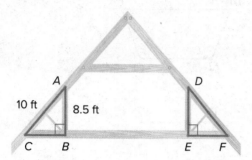

8. In the city park map shown, △DEF ≅ △JKL. The distance from D to E is 20 yards and the distance from D to F is 40 yards. What is the distance from K to L? Round to the nearest tenth.

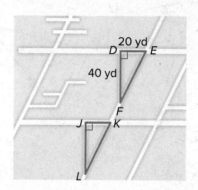

9. **Create** Write a real-world problem involving congruent figures in which you would need to find the measure of a missing angle or side.

10. Determine if the statement is *true* or *false*. Write an argument that can be used to defend your solution.

   If two figures are congruent, then their areas are equal.

11. Determine whether the statement is true or false. Create several pairs of triangles and measure the corresponding sides and angles to justify your response.

   If three sides of one triangle are congruent to the corresponding sides of another triangle, then the two triangles are congruent.

12. MP **Find the Error** A student wrote the congruence statement △NOQ ≅ △DLS for the congruent triangles shown. Find the student's mistake and correct it.

Similarity and Transformations

I Can... determine if two figures are similar by determining a sequence of rotations, reflections, translations, and dilations that maps one similar figure onto another.

What Vocabulary Will You Learn?
similar

Learn Similarity

In a dilation, the scale factor is the ratio of the side lengths of the image to the side lengths of the preimage. When the scale factor is not equal to one, a dilation changes the size of a figure, but does not change the shape of a figure. If the size is changed, the image and the preimage are *not* congruent.

The following are dilations of rectangle *ABCD*.

Scale Factor 0.5	Scale Factor 1.0	Scale Factor 1.5

You can show two figures are **similar** if the second can be obtained from the first by a sequence of dilations and congruence transformations (translations, reflections, rotations).

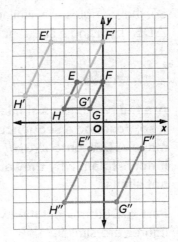

Since a dilation does not change the shape of a figure, the image and the preimage are similar.

On the graph, parallelogram *E′F′G′H′* is the dilated image of parallelogram *EFGH*, and parallelogram *E′F′G′H′* can be mapped onto parallelogram *E″F″G″H″* using a translation 3 units right and 8 units down.

(continued on next page)

◉ Go Online Watch the animation to see how you can use transformations to determine if △ABC is similar to △DEF.

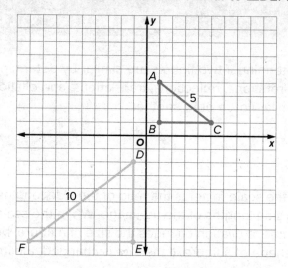

Step 1 Write ratios comparing the lengths of each side.

$\frac{DF}{AC} = \frac{10}{5}$ or $\frac{2}{1}$ $\qquad$ $\frac{EF}{BC} = \frac{8}{4}$ or $\frac{2}{1}$ $\qquad$ $\frac{DE}{AB} = \frac{6}{3}$ or $\frac{2}{1}$

△DEF is the dilated image of △ABC with a scale factor of 2.

Step 2 On the graph above, dilate △ABC with a center of dilation at the origin and a scale factor of 2.

Step 3 Reflect △A′B′C′ across the y-axis.

Step 4 Translate △A″B″C″ down 10 units and to the right 1 unit so that it maps exactly onto △DEF.

So, the triangles are similar because △ABC can be mapped onto △DEF by using a sequence of a dilation, a reflection, and a translation.

(continued on next page)

Example 1 Determine Similarity

Are the two figures similar? If so, describe a sequence that maps △DEF onto △GHI. If not, explain why they are not similar.

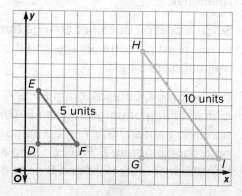

Part A Determine if the two figures are similar.

Step 1 Determine if a dilation occurred by examining the ratios of the side lengths.

$\frac{GH}{DE} = \frac{8}{4}$ or ☐ $\qquad$ $\frac{GI}{DF} = \frac{6}{3}$ or ☐ $\qquad$ $\frac{HI}{EF} = \frac{10}{5}$ or ☐

Since the ratios are equal, a dilation, with a scale factor of 2, is one of the transformations that maps △DEF onto △GHI.

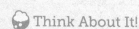

Talk About It!

Compare and contrast using transformations to prove that two triangles are congruent versus using transformations to prove that two triangles are similar.

Think About It!

How can the side measures help you determine if the triangles are similar?

Step 2 Graph the transformations.

Graph the dilation of △DEF with a center of dilation at the origin and a scale factor of 2. Then translate △D'E'F' seven units right and three units down so that △D'E'F' maps onto △GHI.

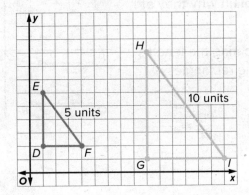

Since △DEF is mapped onto △GHI with a dilation followed by a translation, the two triangles are similar.

Part B Describe the sequence of transformations.

A dilation with center at the origin and a scale factor of _____

followed by a translation _____ units right and _____ units

down maps △DEF onto △GHI.

Check

Are the two figures similar? If so, describe a sequence that maps △STU onto △PQR. If not, explain why they are not similar.

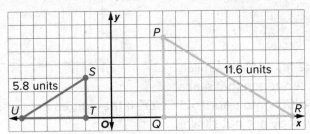

Part A Determine if the figures are similar.

Part B If the figures are similar, describe a sequence that maps △STU onto △PQR. If the figures are not similar, explain why they are not similar.

🌀 **Go Online** You can complete an Extra Example online.

> 💬 **Talk About It!**
> Is a dilation alone sufficient to map △DEF onto △GHI? Explain.

Example 2 Determine Similarity

How would you begin solving this problem?

Is there a sequence of rotations, reflections, translations, and/or dilations that would map rectangle WXYZ onto rectangle RSPQ?

Are the two figures similar? If so, describe a sequence that maps rectangle WXYZ onto rectangle RSPQ. If not, explain why they are not similar.

Write ratios comparing the lengths of each side of the rectangles.

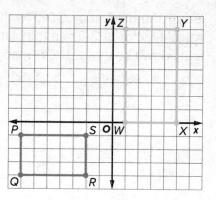

$$\frac{SP}{XY} = \frac{\boxed{}}{\boxed{}} \qquad \frac{PQ}{YZ} = \frac{\boxed{}}{\boxed{}}$$

$$\frac{QR}{ZW} = \frac{\boxed{}}{\boxed{}} \qquad \frac{RS}{WX} = \frac{\boxed{}}{\boxed{}}$$

The ratios are not equivalent. So, the two rectangles are not similar because a dilation did not occur.

Check

Are the two figures similar? If so, describe a sequence that maps trapezoid ABCD onto trapezoid GHEF. If not, explain why they are not similar.

Part A Determine if the figures are similar.

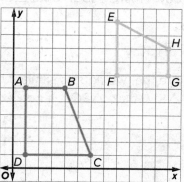

Part B If the figures are similar, describe a sequence that maps trapezoid ABCD onto trapezoid GHEF. If the figures are not similar, explain why they are not similar.

 Go Online You can complete an Extra Example online.

Learn Identify Transformations

Similar figures have the same shape, but may have different sizes. The sizes of the two similar figures are related to the scale factor of the dilation.

If the scale factor of the dilation is ...	then the dilated figure is ...
between 0 and 1	smaller than the original
equal to 1	the same size as the original
greater than 1	larger than the original

You can determine the sequence of a dilation followed by a congruence transformation that maps one figure onto a similar figure.

Example 3 Identify Transformations

Square ABCD is similar to square EFGH. Determine which sequence of transformations maps square ABCD onto square EFGH.

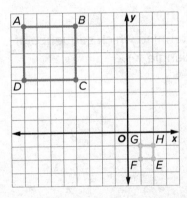

Step 1 Since the figures are similar, a dilation occurred. Find the scale factor of the dilation.

Write the ratios comparing the side lengths.

$$\frac{EF}{AB} = \frac{\boxed{}}{\boxed{}} \qquad \frac{FG}{BC} = \frac{\boxed{}}{\boxed{}}$$

$$\frac{GH}{CD} = \frac{\boxed{}}{\boxed{}} \qquad \frac{HE}{DA} = \frac{\boxed{}}{\boxed{}}$$

The scale factor of the dilation is $\frac{1}{4}$.

(continued on next page)

⬤ Think About It!

Without calculating, is the scale factor of the dilation *between 0 and 1, equal to 1*, or *greater than 1?*

Step 2 Graph the transformations.

Graph the dilation of square *ABCD* with a center of dilation at the origin and scale factor of $\frac{1}{4}$. Then rotate square *A'B'C'D'* 180° clockwise about the origin so that the squares coincide.

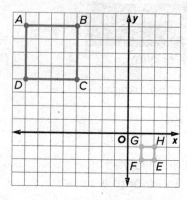

Talk About It!

Are all squares similar? If so, what does that tell you about possible transformations performed?

So, the transformations that map square *ABCD* onto square *EFGH* consist of a dilation with a center at the origin and a scale factor of $\frac{1}{4}$ followed by a 180° clockwise rotation about the origin.

Check

Triangle *ABC* is similar to triangle *XYZ*. Determine which sequence of transformations maps triangle *ABC* onto triangle *XYZ*.

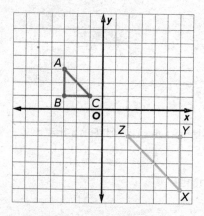

Show your work here

Go Online You can complete an Extra Example online.

🌐 Example 4 Use the Scale Factor

Ken enlarges the photo shown by a scale factor of 2 for his web page. He then enlarges the web page photo by a scale factor of 1.5 to print. The original photo is 2 inches by 3 inches.

What are the dimensions of the print photo? Are the enlarged photos similar to the original?

Part A Find the dimensions.

Multiply each dimension of the original photo by 2 to find the dimensions of the web page photo.

2 in. × 2 = _____ in.

3 in. × 2 = _____ in.

So, the web page photo will be 4 inches by 6 inches.

Multiply the dimensions of that photo by 1.5 to find the dimensions of the print.

4 in. × 1.5 = _____ in.

6 in. × 1.5 = _____ in.

So, the printed photo will be 6 inches by 9 inches.

Part B Determine similarity.

Each enlargement was the result of a _____. If the three photos were placed next to one another, then a dilation followed by a translation maps the photos onto each other.

So, all three photos are similar.

🍄 Think About It!

How would you begin solving the problem?

💬 Talk About It!

Would the dimensions of the print be the same if Ken first dilated the photo by a scale factor of 1.5 and then by a scale factor of 2? Why or why not?

Check

An art show offers different-sized prints of the same painting. The original print measures 24 centimeters by 30 centimeters. A printer enlarges the original by a scale factor of 1.5, and then enlarges the second image by a scale factor of 3.

Part A

What are the dimensions of the largest print?

Part B

Are both of the enlarged prints similar to the original?

Show your work here

 Go Online You can complete an Extra Example online.

Pause and Reflect

Compare and contrast *congruent figures* and *similar figures*. How are they the alike? How are they different?

Record your observations here

🌐 Apply Careers

A designer enlarges an image with a length of 6 centimeters and width of 9 centimeters by a scale factor of 3. The designer decides that the enlarged image is too large and reduces it by a scale factor of 0.5. Will the final image fit into a rectangular space that has an area of 121 square centimeters? Explain your answer.

1 What is the task?

Make sure you understand exactly what question to answer or problem to solve. You may want to read the problem three times. Discuss these questions with a partner.

First Time Describe the context of the problem, in your own words.
Second Time What mathematics do you see in the problem?
Third Time What are you wondering about?

2 How can you approach the task? What strategies can you use?

Record your observations here

3 What is your solution?

Use your strategy to solve the problem.

Show your work here

💬 Talk About It!

How did the properties of similar figures help you solve the problem?

4 How can you show your solution is reasonable?

✏️ **Write About It!** Write an argument that can be used to defend your solution.

Check

An artist enlarges a rectangular painting that has a length of 12 inches and width of 16 inches by a scale factor of 2. He then decides to reduce the enlarged image by a scale factor of 0.4. Will the final painting fit into a rectangular frame that has an area of 120 square inches? Write an argument that can be used to defend your solution.

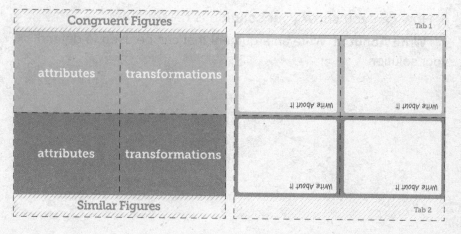

Go Online You can complete an Extra Example online.

Foldables It's time to update your Foldable, located in the Module Review, based on what you learned in this lesson. If you haven't already assembled your Foldable, you can find the instructions on page FL1.

Practice

🔖 **Go Online** You can complete your homework online.

Determine if each pair of figures is similar. If so, describe a sequence of transformations that maps one figure onto the other figure. If not, explain why they are not similar. (Examples 1 and 2)

1.

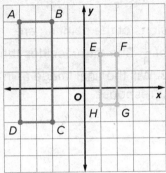

2.

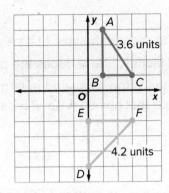

3. Triangle *ABC* is similar to △*XYZ*. Determine which sequence of transformations maps △*ABC* onto △*XYZ*. (Example 3)

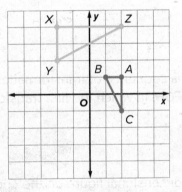

Test Practice

4. Jenna is creating a mural for her bedroom wall. She would like to copy a picture that is 2 inches by 2.5 inches. She uses a copy machine to enlarge it by a scale factor of 4. Then she projects it on her wall by a scale factor of 12. What are the dimensions of the mural? Are the enlarged pictures similar to the original? (Example 4)

5. Multiple Choice Which sequence of transformations can be used to show that two figures are similar but not necessarily congruent?

- Ⓐ dilation and rotation
- Ⓑ translation and reflection
- Ⓒ reflection and rotation
- Ⓓ rotation and translation

Apply

6. A graphic designer enlarges a rectangular image with a length of 3 inches and width of 5 inches by a scale factor of 2. Then he decides that the enlarged image is too large and reduces it by a scale factor of 0.25. Will the final image fit into a rectangular space that has an area of 3.5 square inches? Justify your response.

7. An artist needs to reduce the size of a painting. The original dimensions of the painting are 12 inches by 20 inches. She reduces the painting by a scale factor of $\frac{1}{4}$. She then decides that the reduced image is too small and enlarges it by a scale factor of 2. Will the final image fit in a rectangular space that has an area of 55 square inches? Justify your response.

8. Square ABCD is similar to square EFGH because a dilation with a scale factor of 2 with the center of dilation at the origin, followed by a translation 5 units to the right maps square ABCD onto square EFGH.

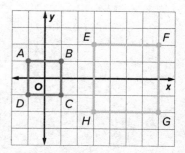

   a. If you perform the translation first and then the dilation, will the squares still map onto one another? Explain.

   b. Describe a sequence of transformations that maps square ABCD onto square EFGH, in which the first transformation is a translation.

9. Draw a two-dimensional figure on the coordinate plane. Then perform a series of transformations on the figure. Which figures are congruent? Which figures are similar?

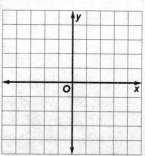

10. (MP) **Find the Error** A student concluded that rectangle ABCD is similar to rectangle EFGH because a dilation with a scale factor of 0.5 and a translation maps rectangle ABCD onto rectangle EFGH. Find the student's mistake and correct it.

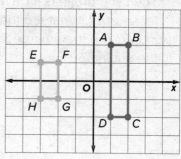

Similarity and Corresponding Parts

I Can... use properties of similar figures to determine similarity and to find missing measures.

Explore Similar Triangles

🖰 **Online Activity** You will use Web Sketchpad to explore properties of similar triangles.

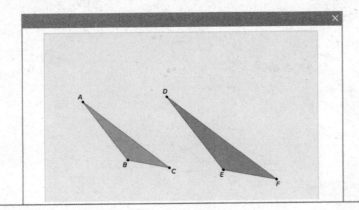

Learn Similar Polygons

Polygons that have the same shape are called *similar polygons*. The parts of similar figures that "match" are called *corresponding parts*. The notation △ABC ~ △XYZ is read *triangle ABC is similar to triangle XYZ*.

Words	Symbols
If two polygons are similar, then their corresponding angles are congruent and the lengths of the corresponding sides are proportional.	△ABC ~ △XYZ Congruent angles: ∠A ≅ ∠X; ∠B ≅ ∠Y; ∠C ≅ ∠Z;
Model	Corresponding sides:
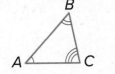	$\frac{AB}{XY} = \frac{BC}{YZ} = \frac{AC}{XZ}$

Example 1 Write Similarity Statements

Determine whether rectangle *HJKL* is similar to rectangle *MNPQ*. If so, write a similarity statement.

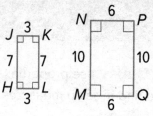

Step 1 Check to see if corresponding angles are congruent.

Since the two polygons are rectangles, all of their angles are _____ angles.
Therefore, all corresponding angles are congruent.

Step 2 Check to see if corresponding side lengths are proportional.

Write the ratios of the corresponding sides in simplest form.

$$\frac{HJ}{MN} = \frac{\Box}{\Box} \qquad\qquad \frac{JK}{NP} = \frac{3}{6} \text{ or } \frac{\Box}{\Box}$$

$$\frac{KL}{PQ} = \frac{\Box}{\Box} \qquad\qquad \frac{LH}{QM} = \frac{3}{6} \text{ or } \frac{\Box}{\Box}$$

Since $\frac{7}{10}$ and $\frac{1}{2}$ are not equivalent, the corresponding sides are not proportional.

So, the rectangles are *not* similar.

💬 **Talk About It!**

Can you assume that the two rectangles are similar just because their corresponding angles are congruent? Explain.

Check

Which of the following is true about △*ABC* and △*XYZ*? Select all that apply.

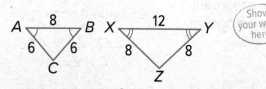

Show your work here

☐ The triangles are similar.

☐ The triangles are not similar.

☐ The triangles are congruent.

☐ △*ABC* ~ △*XYZ*

☐ △*ABC* ≅ △*XYZ*

🔵 **Go Online** You can complete an Extra Example online.

Explore Angle-Angle Similarity

Online Activity You will use Web Sketchpad to explore the angle-angle criterion for triangle similarity.

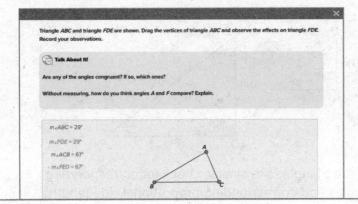

Learn Angle-Angle (AA) Similarity

To show that two triangles are similar, you do not need to show that all of their corresponding angles are congruent and that all of their corresponding side lengths are proportional.

Words
The **Angle-Angle Similarity** states that if two angles of one triangle are congruent to two angles of another triangle, then the triangles are similar.

Model
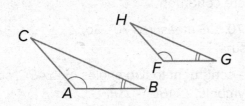

Symbols
If $\angle A \cong \angle F$ and $\angle B \cong \angle G$, then $\triangle ABC \sim \triangle FGH$.

(continued on next page)

Talk About It!

How do you know the third pair of angles in the triangles is congruent?

Go Online Watch the video to see how Angle-Angle Similarity can be demonstrated.

The video shows that since two angles of △ABC are congruent to two angles of △DEF, then △ABC ~ △DEF.

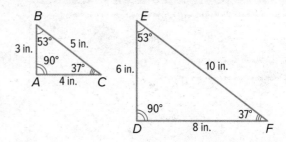

To verify that the triangles are similar, check for corresponding congruent angles and corresponding proportional side lengths.

Are all of the corresponding angles congruent?

Are the measures of the corresponding sides proportional?

Example 2 Angle-Angle Similarity

Determine whether the triangles are similar. If so, write a similarity statement.

Angle A and ∠_____ have the same measure, so they are congruent.

Since 180 − 62 − 48 = 70, ∠G measures 70°. So, ∠_____ and ∠G have the same measure.

Two angles of △EFG are congruent to two angles of △ABC, so the triangles are similar. In symbols, △ABC ~ △EFG.

Check

Which of the following is true about △JKH and △MKL? Select all that apply.

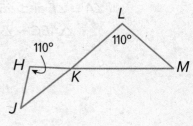

☐ The triangles are similar.

☐ The triangles are not similar.

☐ △JHK ~ △MKL

☐ △JHK ≅ △MKL

☐ ∠J ≅ ∠L

Go Online
You can complete an Extra Example online.

Learn Similar Polygons and Scale Factor

Scale factor is the ratio of the lengths of two corresponding sides of two similar polygons.

Since the corresponding sides are proportional, you can use the scale factor or a proportion to find missing measures.

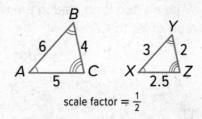

scale factor = $\frac{1}{2}$

Example 3 Find Missing Measures

The quadrilaterals are similar. Find the missing side measure.

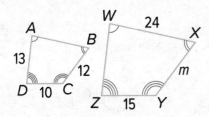

Think About It!

Because the quadrilaterals are similar, what do you know about the corresponding sides?

Method 1 Use the scale factor.

Step 1 Find the scale factor.

Find the scale factor from quadrilateral *ABCD* to quadrilateral *WXYZ*.

Find the ratio of corresponding sides with known lengths.

scale factor: $\dfrac{YZ}{CD} = \dfrac{15}{10}$ or $\dfrac{\boxed{}}{\boxed{}}$

So, a length on quadrilateral *WXYZ* is $\frac{3}{2}$ times as long as the corresponding length on quadrilateral *ABCD*.

Step 2 Write an equation to find the missing measure.

Segments *BC* and *XY* are corresponding parts. Let *m* represent the measure of $\overline{XY}$.

$m = \dfrac{3}{2}\,(12)$ Write the equation.

$m = \boxed{}$ Multiply.

So, the length of the missing side is 18 units.

(continued on next page)

Method 2 Use a proportion.

Write an equation stating the ratio of XY to BC is equivalent to the ratio of YZ to CD.

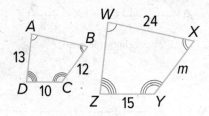

$$\frac{XY}{BC} = \frac{YZ}{CD}$$ 　　Write the proportion.

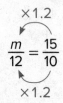

$XY = m$, $BC = 12$, $YZ = 15$, $CD = 10$

Find an equivalent ratio. Because $10 \cdot 1.2 = 12$, multiply $15 \cdot 1.2$ to find m.

$$\frac{\boxed{}}{12} = \frac{15}{10}$$ 　　Multiply.

So, using either method, the length of the missing side is 18 units.

Check

The triangles are similar. Find the missing side measure.

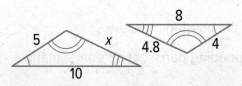

🔎 **Go Online** You can complete an Extra Example online.

🌐 **Apply** Fashion Design

Alex used reflective tape to make the design shown on a jacket. First, he made the small polygon. Then he enlarged the small polygon to make the large polygon, using a scale factor that extended the 8-centimeter side by 2 centimeters. What total length of reflective tape did Alex use?

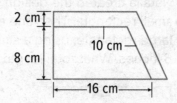

🔎 Go Online
Watch the animation.

1 What is the task?

Make sure you understand exactly what question to answer or problem to solve. You may want to read the problem three times. Discuss these questions with a partner.

First Time Describe the context of the problem, in your own words.
Second Time What mathematics do you see in the problem?
Third Time What are you wondering about?

2 How can you approach the task? What strategies can you use?

3 What is your solution?

Use your strategy to solve the problem.

💬 Talk About It!

How can you use the scale factor to find the dimensions of a third similar polygon?

4 How can you show your solution is reasonable?

🖊 **Write About It!** Write an argument that can be used to defend your solution.

Check

Natalia created the design shown using ribbon. First, she made the small rectangle. Then she enlarged the small rectangle to make the large rectangle, using a scale factor that extended the 2-inch side by 6 inches. What total length of ribbon did Natalia use?

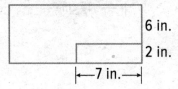
6 in.
2 in.
|← 7 in. →|

Show your work here

Go Online You can complete an Extra Example online.

Foldables It's time to update your Foldable, located in the Module Review, based on what you learned in this lesson. If you haven't already assembled your Foldable, you can find the instructions on page FL1.

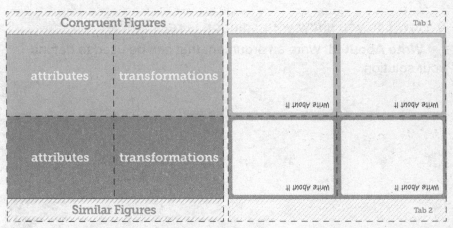

Practice

⬤ **Go Online** You can complete your homework online.

Determine whether each pair of polygons is similar. If so, write a similarity statement. (Examples 1 and 2)

1.

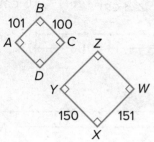

2.

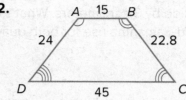

3.

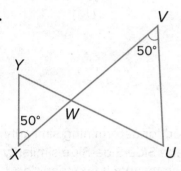

4.

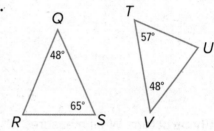

Each pair of polygons is similar. Find each missing side measure. (Example 3)

5.

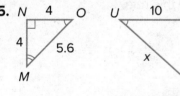

6.

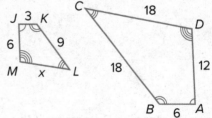

Test Practice

7. Multiselect Which of the following is true about △ABC and △XYZ? Select all that apply.

☐ The triangles are similar.

☐ The triangles are not similar.

☐ The triangles are congruent.

☐ △ABC ~ △XYZ

☐ △ABC ≅ △XYZ

8. Samantha used craft wire to make the design shown. She first made the smaller quadrilateral. Then she enlarged the smaller quadrilateral to make the larger quadrilateral, using a scale factor that extended the 6-centimeter side by 3 centimeters. What total length of craft wire did Samantha use for both quadrilaterals?

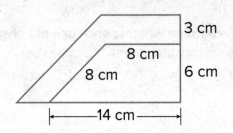

9. Triangle *JKL* is reduced to obtain triangle *MNL*. Using the same scale factor, triangle *MNL* is reduced to obtain triangle *OPL*. If *JK* = 35 millimeters, *KL* = 25 millimeters, *LJ* = 25 millimeters, and *MN* = 28 millimeters, what is the perimeter of triangle *OPL*?

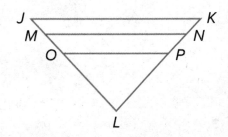

10. (MP) **Identify Structure** What measures must be known in order to find the missing side measure of a pair of similar quadrilaterals?

11. Another method for determining similarity of triangles is the Side-Side-Side similarity rule. This rule states that if two triangles have three pairs of corresponding sides in the same ratio, then the triangles are similar. Create several pairs of triangles and measure the corresponding sides and angles to verify this rule.

12. (MP) **Reason Abstractly** Determine if the statement is *true* or *false*. Justify your response.

If two figures are similar, then their perimeters are related by the scale factor of the figures.

13. A blue rectangular tile and a red rectangular tile are similar. The blue tile has a length of 10 inches and a perimeter of 30 inches. The red tile has a length of 6 inches. What is the perimeter of the red tile?

Indirect Measurement

I Can... use properties of similar triangles to solve indirect measurement problems.

Explore Similar Triangles and Indirect Measurement

Online Activity You will use Web Sketchpad to explore problems involving similar triangles.

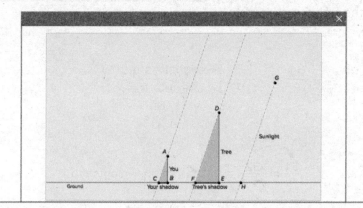

Learn Indirect Measurement

Indirect measurement allows you to use properties of similar polygons to find distances or lengths that are difficult to measure directly.

One type of indirect measurement is *shadow reckoning*. Two objects and their shadows form two sides of right triangles. In shadow problems, you can assume that the angles formed by the Sun's rays with two objects at the same location are congruent. Since two pairs of corresponding angles are congruent, the two right triangles are similar.

What are the corresponding sides of the right triangles?

 Talk About It!

Is there another proportion that can be used to solve the problem?

 Talk About It!

Why are the right triangles in the diagram similar?

 **Example 1** Use Indirect Measurement

The lead scout statue of the Korean War Memorial in Washington, D.C., casts a 43.5-inch shadow at the same time a nearby tourist casts a 32-inch shadow.

If the tourist is 64 inches tall, how tall is the statue?

Write a proportion comparing the shadow lengths and heights. Let h represent the unknown height of the statue.

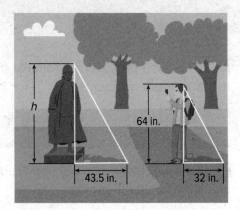

tourist's height → $\dfrac{64}{h} = \dfrac{32}{43.5}$ ← tourist's shadow
statue's height →   ← statue's shadow

$$\dfrac{64}{h} = \dfrac{32}{43.5}$$ Find an equivalent ratio.

$$\dfrac{64}{87} = \dfrac{32}{43.5}$$ Multiply 43.5 by 2.

So, $h = 87$. The statue is 87 inches tall.

Check

How tall is the tree?

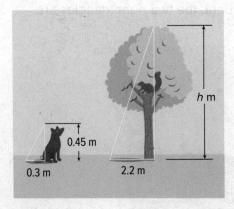

 **Go Online** You can complete an Extra Example online.

Example 2 Use Indirect Measurement

In the figure, △DBA is similar to △ECA.

Find the distance _d_ across the lake.

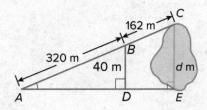

Separate the triangles as shown.

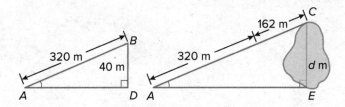

Write a proportion to find the missing measure.

$$\frac{AB}{AC} = \frac{BD}{CE}$$ $\overline{AB}$ corresponds to $\overline{AC}$, and $\overline{BD}$
                  corresponds to $\overline{CE}$.

$$\frac{\boxed{}}{\boxed{}} = \frac{\boxed{}}{\boxed{}}$$ $AB = 320;\ AC = 482;$
                  $BD = 40;\ CE = d$

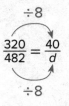

$$\frac{320}{482} = \frac{40}{d}$$ Find an equivalent ratio.

÷8

$$\frac{320}{482} = \frac{40}{60.25}$$ Divide 482 by 8.

So, _d_ = 60.25. The distance across the lake is 60.25 meters.

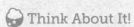

Think About It!
What is true about the
corresponding sides of
similar triangles?

Check

In the figure, $\triangle ABC$ is similar to $\triangle EBD$. Find the distance x across the ravine.

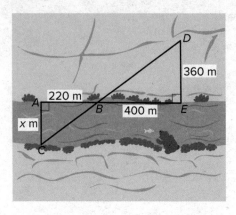

Show your work here

🖱 **Go Online** You can complete an Extra Example online.

Pause and Reflect

How can you use similarity to solve real-world problems? Include an example in your explanation.

Record your observations here

Practice

🔅 **Go Online** You can complete your homework online.

1. Becky casts a 7-foot shadow at the same time a nearby mailbox casts a 4-foot shadow. If the mailbox is 3 feet tall, how tall is Becky? (Example 1)

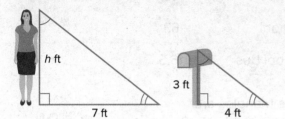

h ft

3 ft

7 ft   4 ft

2. At the same time a $6\frac{1}{2}$-foot tall teacher casts a 9-foot shadow, a nearby flagpole casts a $31\frac{1}{2}$-foot shadow. How tall is the flagpole? (Example 1)

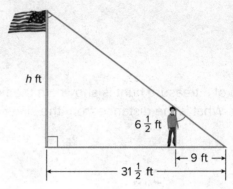

h ft

$6\frac{1}{2}$ ft

← 9 ft →

$31\frac{1}{2}$ ft

3. In the figure, △ABE is similar to △ACD. What is the height h of the ramp when it is 2 feet from the building? (Example 2)

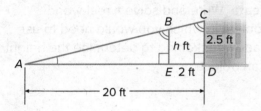

B   C

h ft   2.5 ft

A

E 2 ft D

20 ft

4. In the figure, the triangles are similar. What is the distance d from the water ride to the roller coaster? Round to the nearest tenth. (Example 2)

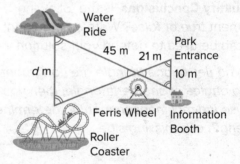

Water Ride

45 m   21 m   Park Entrance

d m   10 m

Ferris Wheel   Information Booth

Roller Coaster

Test Practice

5. If a 25-foot-tall house casts a 75-foot shadow at the same time that a streetlight casts a 60-foot shadow, how tall is the streetlight?

6. Table Item A child and a statue casts the shadow lengths shown at the same time. Complete the table to find the height, in feet, of the statue.

Object	Height of Object (ft)	Shadow Length (ft)
Emma	3.5	5.25
Statue		57

Apply

7. Mr. Nolan's math class went out to measure shadows in their school yard. Their data is recorded in the table. Find the missing heights.

Person/Item	Shadow Length (ft)	Height of Person/Item (ft)
Mr. Nolan	9	6
Flagpole	48	
School	63	
School Bus	16.5	

8. A map of a treasure hunt is shown. In the figure, the triangles are similar. What is the distance from the silver coins to the gold coins?

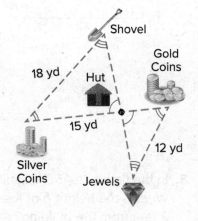

9. (MP) **Justify Conclusions** Is the following statement *true or false*? Write an argument that can be used to defend your solution.

If two angles of one triangle are congruent to two angles of another triangle, then you can use indirect measurement to determine the length of a missing side.

10. Create Write and solve a real-world problem in which you would need to use shadow reckoning to determine the height of an object.

11. (MP) **Find the Error** A student used the proportion below to find the person's height *h* shown in the diagram. Find the student's mistake and correct it.

$$\frac{h}{5} = \frac{20}{25}$$

$$h = 4$$

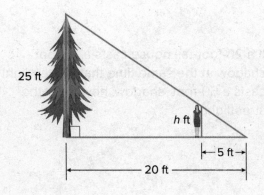

⬡ **Foldables** Use your Foldable to help review the module.

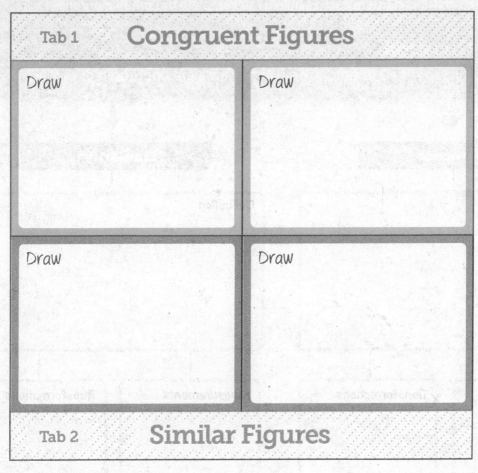

Tab 1 — **Congruent Figures**

Draw

Draw

Draw

Draw

Tab 2 — **Similar Figures**

Rate Yourself! ◼ ◆ ★

Complete the chart at the beginning of the module by placing a checkmark in each row that corresponds with how much you know about each topic after completing this module.

Write about one thing you learned.

_____

_____

_____

_____

_____

_____

Write about a question you still have.

_____

_____

_____

_____

_____

_____

Reflect on the Module

Use what you learned about congruence and similarity to complete the graphic organizer.

e Essential Question

What information is needed to determine if two figures are congruent or similar?

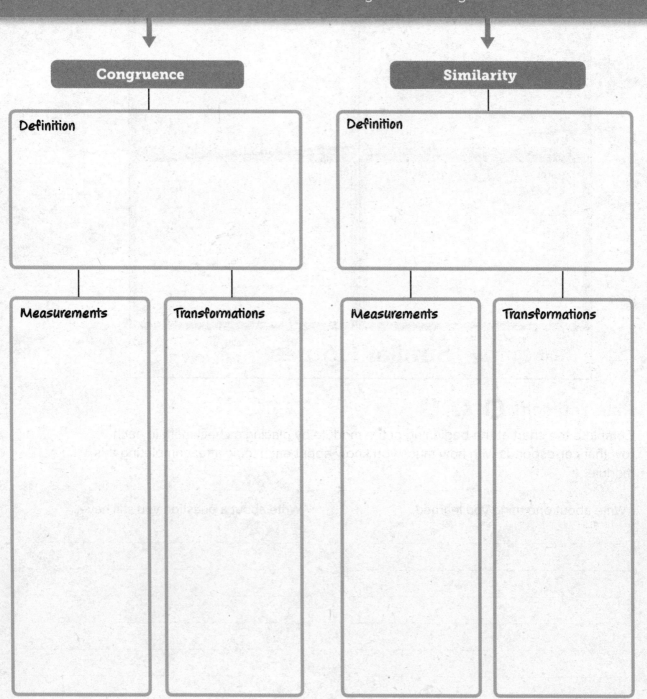

Congruence

Definition

Measurements

Transformations

Similarity

Definition

Measurements

Transformations

Test Practice

1. Open Response Consider quadrilaterals *ABCD* and *JKLM* as shown. (Lesson 1)

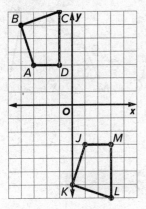

A. Are the quadrilaterals congruent?

B. If the quadrilaterals are congruent, describe the sequence of transformations that maps quadrilateral *ABCD* onto quadrilateral *JKLM*. If the figures are not congruent, explain why they are not congruent.

2. Multiple Choice Triangle *ABC* is congruent to △*PQR*. Which of the following sequence of transformations maps △*ABC* onto △*PQR*? (Lesson 1)

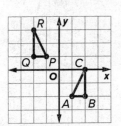

Ⓐ reflection across *y*-axis, translation 3 units up

Ⓑ reflection across *x*-axis, translation 3 units up

Ⓒ translation 2 units left, translation 3 units up

Ⓓ rotation 90° clockwise about the origin, translation 3 units up

3. Table Item Consider quadrilaterals *ABCD* and *WXYZ* as shown. (Lesson 2)

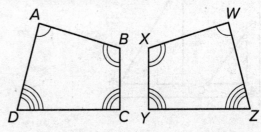

Indicate whether each of the following statements is correct or incorrect regarding the congruency of these quadrilaterals.

Statements	Correct	Incorrect
$\overline{AB} \cong \overline{WX}$		
$\overline{BC} \cong \overline{WZ}$		
$\overline{AD} \cong \overline{XY}$		
∠*B* ≅ ∠*X*		
∠*A* ≅ ∠*W*		
∠*C* ≅ ∠*Z*		

4. Multiple Choice A diagram of a truss bridge is shown. In the diagram △*ABC* ≅ △*ADC*. If *AC* = 32 feet and *DC* = 26 feet, what is the length of $\overline{AB}$? Round to the nearest tenth. (Lesson 2)

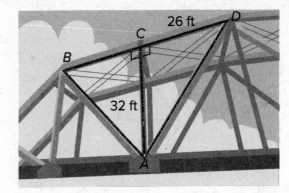

Ⓐ 18.7 feet

Ⓑ 36.5 feet

Ⓒ 38.4 feet

Ⓓ 41.2 feet

5. Open Response Consider triangles *ABC* and *JKL* as shown. (Lesson 3)

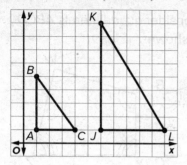

A. Are the triangles similar?

B. If the triangles are similar, describe the sequence of transformations that maps triangle *ABC* onto triangle *JKL*. If the figures are not similar, explain why they are not similar.

6. Multiselect Select all the statements that accurately describe the triangles shown.
(Lesson 4)

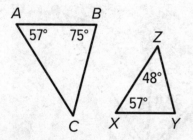

☐ $m\angle C = 48°$

☐ $m\angle Y = 75°$

☐ $m\angle C = m\angle X$

☐ $\triangle ABC \sim \triangle XYZ$

☐ $\triangle ABC \cong \triangle XXZ$

7. Equation Editor The quadrilaterals shown are similar. Find the missing side measure, *x*, in units. (Lesson 4)

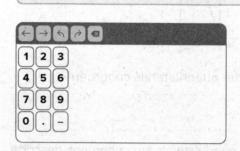

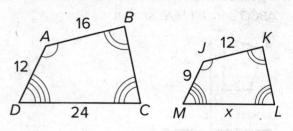

8. Equation Editor A tree in the school yard casts a 60-inch shadow at the same time a nearby student casts a 17-inch shadow. If the student is 68 inches tall, how many <u>feet</u> tall is the tree? (Lesson 5)

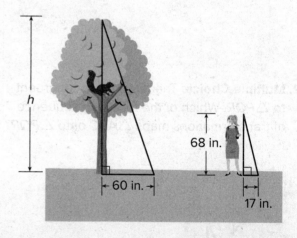

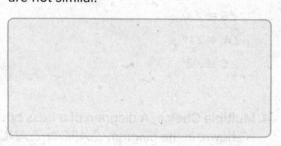

e Essential Question

How can you measure a cylinder, cone, or sphere?

What Will You Learn?

Place a checkmark (✓) in each row that corresponds with how much you already know about each topic **before** starting this module.

KEY	Before			After		
⬛ — I don't know. ◆ — I've heard of it. ★ — I know it!	⬛	◆	★	⬛	◆	★
finding volume of cylinders						
finding volume of cones						
finding volume of spheres and hemispheres						
finding missing dimensions of cylinders, cones, and spheres, given the volume and the other dimension(s) of the figure						
finding volume of composite solids						

📖 Foldables Cut out the Foldable and tape it to the Module Review at the end of the module. You can use the Foldable throughout the module as you learn about volume.

What Vocabulary Will You Learn?

Check the box next to each vocabulary term that you may already know.

- ☐ composite solids
- ☐ cone
- ☐ cylinder
- ☐ hemisphere
- ☐ sphere
- ☐ volume

Are You Ready?

Study the Quick Review to see if you are ready to start this module.

Then complete the Quick Check.

Quick Review

Example 1

Find circumference of circles.

Find the circumference of a circle with a diameter of 12 feet. Use 3.14 for π.

$C = \pi d$	Circumference of a circle
$C \approx 3.14(12)$	Replace π with 3.14 and d with 12.
$C \approx 37.68$	Simplify.

The circumference is about 37.68 feet.

Example 2

Find the area of circles.

Find the area of a circle with a radius of 5 meters. Use 3.14 for π.

$A = \pi r^2$	Area of a circle
$A \approx 3.14(5)^2$	Replace π with 3.14 and r with 5.
$A \approx 78.5$	Simplify.

The area is about 78.5 square meters.

Quick Check

1. The diameter of a circle is 17 inches. What is the circumference of the circle? Use 3.14 for π.

2. The radius of a circular pizza is 7 inches. What is the area of the pizza? Use 3.14 for π.

How Did You Do?

Which exercises did you answer correctly in the Quick Check?
Shade those exercise numbers at the right.

① ②

Volume of Cylinders

I Can... use the formula for the volume of a cylinder to find the volume of a cylinder given its diameter or radius and the height.

Explore Volume of Cylinders

🧭 **Online Activity** You will explore how calculating the volume of a prism is related to calculating the volume of a cylinder, and then make a conjecture about the formula for the volume of a cylinder.

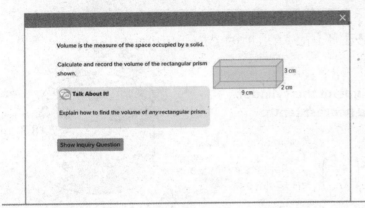

Volume is the measure of the space occupied by a solid.

Calculate and record the volume of the rectangular prism shown.

3 cm
2 cm
9 cm

💬 **Talk About It!**

Explain how to find the volume of *any* rectangular prism.

Show Inquiry Question

Learn Volume of Cylinders

Volume is the measure of the space occupied by a solid. Volume is measured in cubic units. A **cylinder** is a three-dimensional figure with two parallel congruent circular bases connected by a curved surface. The area of the base of a cylinder tells the number of cubic units in one layer. The height tells how many layers there are in the cylinder.

Words	Model
The volume V of a cylinder with radius r is the area of the base B times the height h.	r
Symbols	h
$V = Bh$, where $B = \pi r^2$ or $V = \pi r^2 h$	$B = \pi r^2$

(continued on next page)

Talk About It!

Which of these do you think is a more accurate representation of the cylinder's volume? Explain. What are some advantages and disadvantages to each representation of π?

When solving problems that involve π, you can use the value of π as stored in a calculator, or you can record your answer in terms of π. For example, the volume of the cylinder shown can be represented in either of these ways.

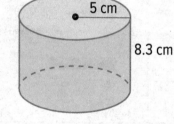

Record Volume in Terms of π	Use the Value of π From a Calculator
$V = \pi r^2 h$	$V = \pi r^2 h$
$V = \pi(4)^2(7)$	$V = \pi(4)^2(7)$
$V = 112\pi \text{ in}^3$	$V \approx 351.9 \text{ in}^3$

Example 1 Find Volume of Cylinders Given the Radius

Find the volume of the cylinder. Round to the nearest tenth.

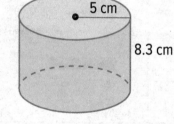

$V = \pi r^2 h$ Volume of a cylinder

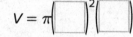   Replace r with 5 and h with 8.3.

     Multiply.

$V \approx$ [ ]   Use a calculator.

So, the volume of the cylinder is about 651.9 cubic centimeters.

Check

Find the volume of the cylinder. Round to the nearest tenth.

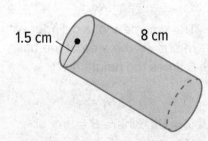

Talk About It!

Why is the term *about* used in the answer? Had you answered in terms of π, would *about* have been necessary?

 Go Online You can complete an Extra Example online.

Example 2 Find Volume of Cylinders Given the Diameter

Find the volume of a cylinder with a diameter of 16 inches and a height of 20 inches. Express your answer in terms of π.

The diameter of a cylinder is 16 inches. The height of the cylinder is 20 inches. Since you are given the diameter, first find the radius. The radius is 8 inches.

$V = \pi r^2 h$ Volume of a cylinder

$V = \pi \boxed{}^2 \boxed{}$ Replace r and h.

$V = \boxed{}\pi$ Multiply and simplify.

So, the volume of the cylinder is 1,280π cubic inches.

Check

Find the volume of a cylinder with a diameter of 8 inches and a height of 8 inches. Express your answer in terms of π.

Show your work here

 Think About It!

What is the relationship between a cylinder's diameter and radius?

 Talk About It!

Why is 1,280π considered the exact volume of the cylinder?

Go Online You can complete an Extra Example online.

Pause and Reflect

Did you make any errors when completing the Check exercise? Describe a method you can use to check your answer.

Record your observations here

Talk About It!

How would the weight of the paperweight change if its height is doubled? its radius?

✦ **Example 3** Solve Problems Involving the Volume of Cylinders

A metal paperweight is in the shape of a cylinder. The paperweight has a height of 1.5 inches and a diameter of 2 inches.

How much does the paperweight weigh if 1 cubic inch of metal weighs 1.8 ounces? Round to the nearest tenth.

Step 1 Find the volume of the paperweight.

$V = \pi r^2 h$ Volume of a cylinder

$V = \pi \left(\boxed{}\right)^2 \left(\boxed{}\right)$ Replace r and h.

$V = \boxed{}\,\pi$ Multiply.

$V \approx \boxed{}$ Use a calculator. Round to the nearest tenth.

The volume of the paperweight is about 4.7 cubic inches.

Step 2 Use the volume to find the weight of the paperweight.

Each cubic inch of metal weighs 1.8 ounces. Multiply the volume, 4.7 cubic inches, by 1.8 to find the weight of the paperweight.

$$4.7(1.8) = \boxed{}$$

So, the weight of the paperweight is about 8.5 ounces.

Check

A scented candle is in the shape of a cylinder. The radius is 4 centimeters and the height is 12 centimeters. Find the mass of the wax needed to make the candle if 1 cubic centimeter of wax has a mass of 3.5 grams. Round to the nearest tenth.

Show your work here

🌐 **Go Online** You can complete an Extra Example online.

🌐 Apply Swimming

A pool with dimensions as shown is filling with water at a rate of 20 gallons per minute. About how many hours will it take to fill the pool if 1 cubic foot of water is about 7.5 gallons? Round to the nearest tenth.

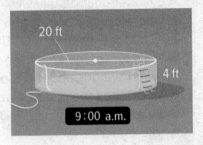

20 ft

4 ft

9:00 a.m.

📷 Go Online
Watch the animation.

1 What is the task?

Make sure you understand exactly what question to answer or problem to solve. You may want to read the problem three times. Discuss these questions with a partner.

First Time Describe the context of the problem, in your own words.
Second Time What mathematics do you see in the problem?
Third Time What are you wondering about?

2 How can you approach the task? What strategies can you use?

Record your observations here

3 What is your solution?

Use your strategy to solve the problem.

Show your work here

💬 Talk About It!

How would doubling the diameter of the pool affect the amount of time it would take to fill? Explain your reasoning.

4 How can you show your solution is reasonable?

✍ **Write About It!** Write an argument that can be used to defend your solution.

Math History Minute

Maryam Mirzakhani (1977–2017) only became interested in mathematics when she was in her last year of high school. In 2014, she became the first woman and the first Iranian honored with the Fields Medal, for her work on hyperbolic geometry. Hyperbolic geometry is used to explore concepts of space and time. The Fields Medal is the highest scientific award for mathematicians and is only presented every four years.

Check

A cylinder-shaped glass with a base radius of 1.5 inches and a height of 6 inches weighs 1.06 ounces when empty. The glass is then filled with water to one inch from the top. If 1 cubic inch of water weighs about 0.6 ounce, how many ounces does the glass of water weigh, including the weight of the glass? Round to the nearest whole ounce.

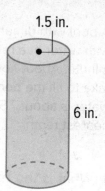

1.5 in.

6 in.

⬩ **Go Online** You can complete an Extra Example online.

Foldables It's time to update your Foldable, located in the Module Review, based on what you learned in this lesson. If you haven't already assembled your Foldable, you can find the instructions on page FL1.

Practice

Go Online You can complete your homework online.

Find the volume of each cylinder. Round to the nearest tenth. (Example 1)

1.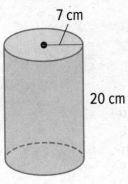
7 cm

20 cm

2.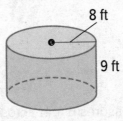
8 ft

9 ft

Find the volume of each cylinder. Express your answer in terms of π. (Example 2)

3.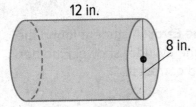
12 in.

8 in.

4.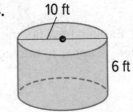
10 ft

6 ft

5. A wooden toy block is in the shape of a cylinder. The toy block has a height of 4 inches and a diameter of 3 inches. How much does the toy block weigh if 1 cubic inch of wood weighs 0.55 ounce? Round to the nearest tenth. (Example 3)

Test Practice

6. A large rainwater collection tub is shaped like a cylinder. The diameter is 28 inches and the height is 40 inches. If the tub is 75% filled, what is the volume of water in the tub? Round to the nearest tenth.

7. Multiple Choice What is the volume of the cylinder shown? (Use 3.14 for π.)

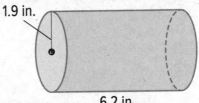
1.9 in.

6.2 in.

Ⓐ 22.382 in³

Ⓑ 70.279 in³

Ⓒ 73.036 in³

Ⓓ 229.333 in³

Apply

8. A soup can, shaped like cylinder, has a diameter of 3.5 inches and a height of 5 inches. Each serving of soup is 15 cubic inches. If a can of soup this size costs $1.99, what is the cost for each serving of soup? Round to the nearest cent.

9. A large water tank measures 6 feet across and 4 feet high. It is being filled with water at a rate of 10 gallons per minute. About how many hours will it take to fill the pool if 1 cubic foot of water is about 7.5 gallons? Round to the nearest tenth.

10. MP Identify Structure Explain how finding the volume of a cylinder is similar to finding the volume of a prism.

11. MP Find the Error A student found the volume of the cylinder shown. Find her mistake and correct it.

$V = \pi r^2 h$

$V = \pi(8^2)(23)$

$V \approx 4{,}624.4 \text{ in}^3$

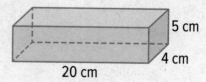

23 in.

8 in.

12. Draw and label a cylinder that has a volume of $1{,}600\pi$ cubic feet.

13. MP Persevere with Problems A cylinder with a height of 17 centimeters and a radius of 8 centimeters is filled with water. If the water is then poured into the rectangular prism shown, will it overflow? Write an argument that can be used to defend your solution.

5 cm

4 cm

20 cm

Volume of Cones

I Can... use the formula for the volume of a cone to find the volume of a cone, given its radius or diameter, and the height.

What Vocabulary Will You Learn?
cone

Explore Volume of Cones

Online Activity You will explore the relationship between the volumes of cones and the volume of cylinders.

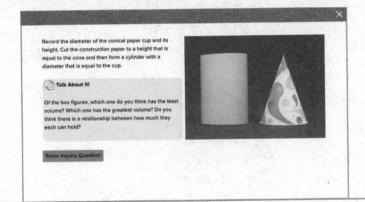

Pause and Reflect

Did you struggle with any of the concepts in this Explore? How do you feel when you struggle with math concepts? What steps can you take to understand those concepts?

Record your observations here

Talk About It!

Compare and contrast the volume formulas for cones and cylinders.

Learn Volume of Cones

A **cone** is a three-dimensional figure with one circular base connected by a curved surface to a single point, called the apex.

In the Explore activity, you learned that there is a relationship between the volume of a cone and cylinder with the same base area and height. The volume of a cone is one-third the volume of a cylinder with the same base area and height.

Words
The volume V of a cone with radius r is one-third the area of the base B times the height h.
Symbols
$V = \frac{1}{3}Bh$ or $V = \frac{1}{3}\pi r^2 h$
Model

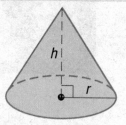

Pause and Reflect

Suppose a cylinder and a cone each have a radius of 5 centimeters. The cylinder and cone each have the same height. If the volume of the cone is approximately 314 cubic centimeters, what is the approximate volume of the cylinder? Write an argument to justify your solution.

Record your observations here

Example 1 Find Volume of Cones

Find the volume of the cone. Express your answer in terms of π.

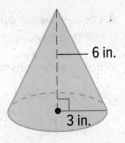

6 in.

3 in.

Since you are given the radius and the height, use the volume formula $V = \frac{1}{3}\pi r^2 h$.

$V = \frac{1}{3}\pi r^2 h$ Volume of a cone

$V = \frac{1}{3}\pi \left(\boxed{}\right)^2 \left(\boxed{}\right)$ Replace r with 3 and h with 6.

$V = \frac{1}{3}\pi \left(\boxed{}\right)$ Multiply.

$V = \boxed{}\ \pi$ Simplify.

So, the volume of the cone is 18π cubic inches.

Check

Find the volume of the cone. Express your answer in terms of π.

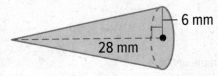

6 mm

28 mm

Show your work here

Lesson 10-2 · Volume of Cones   **545**

Think About It!

Based on the dimensions given, which formula for the volume of a cone is more efficient to use? Why?

Talk About It!

Suppose you forgot the formula for the volume of a cone. How can you use reasoning to find the volume?

Example 2 Find Volume of Cones

A cone-shaped paper cup is filled with water. The height of the cup is 9 centimeters and the diameter is 8 centimeters.

What is the volume of the paper cup? Round to the nearest tenth.

$V = \frac{1}{3}\pi r^2 h$ Volume of a cone

$V = \frac{1}{3}\pi \left(\boxed{}\right)^2 \left(\boxed{}\right)$ Replace r and h.

$V = \boxed{}\ \pi$ Multiply.

$V \approx \boxed{}$ Use a calculator. Round to the nearest tenth.

So, the volume of the paper cup is about 150.8 cubic centimeters.

Check

Find the volume of a cone with a radius of 1.5 inches and a height of 9 inches. Round to the nearest tenth.

Show your work here

Go Online You can complete an Extra Example online.

Pause and Reflect

What part(s) of finding the volume of cones did you feel most confident? Why?

Record your observations here

Think About It!

What dimensions of the cone are needed to solve the problem?

Talk About It!

Suppose a smaller conical paper cup has a diameter of 5 centimeters, and a height of 8 centimeters. Compare and contrast the volumes of the cones using 3.14 for π versus using the π button on a calculator, and rounding the volume to the nearest tenth.

🌏 Apply Popcorn

A family-owned movie theater offers popcorn in the sizes shown. Their cost for the popcorn is $0.09 per cubic inch. If each container is filled to the top, what is the difference between the costs of the popcorn in the two containers?

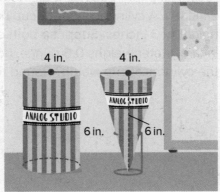

1 What is the task?

Make sure you understand exactly what question to answer or problem to solve. You may want to read the problem three times. Discuss these questions with a partner.

First Time Describe the context of the problem, in your own words.
Second Time What mathematics do you see in the problem?
Third Time What are you wondering about?

2 How can you approach the task? What strategies can you use?

3 What is your solution?

Use your strategy to solve the problem.

4 How can you show your solution is reasonable?

✏️ Write About It! Write an argument that can be used to defend your solution.

💬 Talk About It!

What do you notice about the relationship between the cost of a cylindrical container and the cost of a conical container?

Check

A conical paper cup has a diameter of 3 inches and a height of 3 inches. A cylindrical paper cup has a radius of 1.5 inches and a height of 3 inches. Suppose both cups are filled with water. If 1 cubic inch of water weighs 0.6 ounce, how much more does the water in the cylindrical cup weigh? Round to the nearest tenth.

Go Online You can complete an Extra Example online.

Foldables It's time to update your Foldable, located in the Module Review, based on what you learned in this lesson. If you haven't already assembled your Foldable, you can find the instructions on page FL1.

Practice

Go Online You can complete your homework online.

Find the volume of each cone. Express your answer in terms of π. (Example 1)

1.

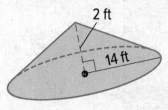

2 ft

14 ft

2.

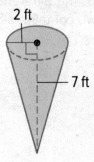

2 ft

7 ft

3.

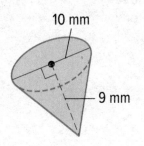

10 mm

9 mm

4.

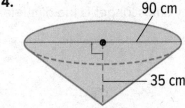

90 cm

35 cm

5. A funnel is in the shape of a cone. The radius is 2 inches and the height is 4.6 inches. What is the volume of the funnel? Round to the nearest tenth. (Example 2)

6. Marta bought a paperweight in the shape of a cone. The radius was 10 centimeters and the height 9 centimeters. Find the volume. Round to the nearest tenth. (Example 2)

Test Practice

7. A lampshade is in the shape of a cone. The diameter is 5 inches and the height is 6.5 inches. Find the volume. Round to the nearest tenth. (Example 2)

8. Multiple Choice What is the volume of the cone shown? (Use 3.14 for π.)

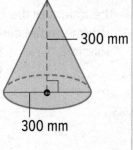

300 mm

300 mm

Ⓐ 7,068,583.5 mm³

Ⓑ 14,137,166.9 mm³

Ⓒ 21,205,750.4 mm³

Ⓓ 229.33304 mm³

Apply

9. A frozen yogurt shop offers frozen yogurt in the sizes shown. The cost per cubic inch is $0.10 for each container's contents. What is the difference between the costs of yogurt in the two containers if each is filled with yogurt?

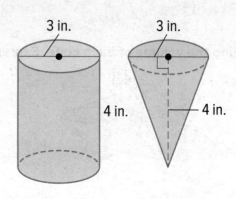

3 in.   3 in.

4 in.   4 in.

10. Cone A and Cone B both have a height of 5 inches. The volume of Cone A is 20.9 cubic inches. The volume of Cone B is 4 times the volume of Cone A. About how many times longer is the diameter of Cone B than the diameter of Cone A?

11. Without calculating, which cone has a greater volume: one with a height of 6 inches and radius of 4 inches or one with a height of 4 inches and radius of 6 inches?

12. Find the volume of the cone with a height of 8 centimeters and a circumference of 18.84 centimeters. Round to the nearest tenth.

13. (MP) **Justify Conclusions** The volumes of a cylinder and a cone are equal. How many times greater is the height of the cone than the height of the cylinder? Write an argument that can be used to defend your solution.

14. (MP) **Find the Error** A student found the volume of the cone shown. Find his mistake and correct it.

$$V = \frac{1}{3}\pi r^2 h$$
$$V = \frac{1}{3}\pi(6^2)(5)$$
$$V = 188.5 \text{ mm}^3$$

5 mm

3 mm

Volume of Spheres

I Can... use the formula for the volume of a sphere or hemisphere to find the volume of the figure given its radius or diameter.

What Vocabulary Will You Learn?
hemisphere

sphere

Learn Volume of Spheres

A **sphere** is a set of all points in space that are a given distance, known as the radius, from a given point, known as the center.

Words	Model
The volume V of a sphere is four-thirds the product of π and the cube of the radius r.	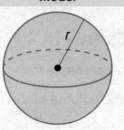
Symbols	
$V = \frac{4}{3}\pi r^3$	

Example 1 Find Volume of Spheres

Find the volume of the sphere. Express your answer in terms of π.

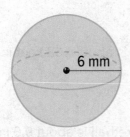

6 mm

$V = \frac{4}{3}\pi r^3$ Volume of a sphere

$V = \frac{4}{3}\pi \left(\boxed{}\right)^3$ Replace r.

$V = \boxed{}\pi$ Multiply.

So, the volume of the sphere is 288π cubic millimeters.

Talk About It!

What do you notice about the relationship between the volume of a sphere with a radius of one inch, and the volume of a sphere with a radius of two inches? Do you think this relationship always exists when a radius is doubled? Explain.

Radius of 1 Inch

$V = \frac{4}{3}\pi r^3$

$V = \frac{4}{3}\pi(1)^3$

$V = \frac{4}{3}\pi$ in^3

Radius of 2 Inches

$V = \frac{4}{3}\pi r^3$

$V = \frac{4}{3}\pi(2)^3$

$V = \frac{32}{3}\pi$ in^3

Check

Find the volume of the sphere.
Express your answer in terms of π.

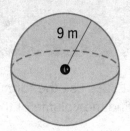

9 m

Show
your work
here

 Example 2 Find Volume of Spheres

A spherical stone found in Costa Rica has a diameter of about 8 feet.

Find the volume of the spherical stone. Round to the nearest tenth.

$V = \frac{4}{3}\pi r^3$ Volume of a sphere

$V = \frac{4}{3}\pi\left(\boxed{}\right)^3$ Replace r.

$V = \boxed{}\,\pi$ Multiply.

$V \approx \boxed{}$ Use a calculator.

So, the volume of the sphere is about 268.1 cubic feet.

Check

The diameter of a men's basketball is about 9.6 inches. What is the volume of a basketball? Round to the nearest tenth.

Show
your work
here

Think About It!

What is the relationship between the radius and the diameter of the sphere?

Talk About It!

Why, in the case of an actual spherical object made of stone, or another material, is it best to report the volume as an approximation and not in terms of π?

 Go Online You can complete an Extra Example online.

🌐 Example 3 Find Volume of Spheres

A training volleyball has a diameter of 10 inches. A pump can inflate the ball at a rate of 325 cubic inches per minute.

How long will it take to inflate the ball?

Step 1 Find the volume of the volleyball. Round to the nearest tenth.

The volume of the volleyball is $\frac{4}{3}\pi(5)^3$ cubic inches, which is about

_____ cubic inches.

Step 2 Write and solve a proportion to find the time to inflate the volleyball.

Round the volume to the nearest tenth.

rate the pump can ⟶ $\dfrac{325 \text{ in}^3}{1 \text{ min}} \approx \dfrac{523.6 \text{ in}^3}{x \text{ min}}$ ⟵ the volume of the ball
inflate the ball   ⟵ time to inflate

Use equivalent ratios to solve the proportion.

$$\frac{325}{1} \approx \frac{523.6}{x} \qquad \text{Write the proportion.}$$

$$\frac{325}{1} \approx \frac{523.6}{x} \qquad \begin{array}{l}\text{Because } 523.6 \div 325 \approx 1.6,\\ \text{multiply 1 by 1.6 to find } x.\end{array}$$

$$\times 1.6 \searrow$$

$$\frac{325}{1} \approx \frac{523.6}{1.6} \qquad \text{Multiply.}$$

$$\times 1.6$$

So, it takes about 1.6 minutes or 1 minute and 36 seconds to inflate the volleyball.

Check

Sarah is blowing up spherical balloons for her brother's birthday party. One of the balloons has a radius of 3 inches. Suppose Sarah can inflate the balloon at a rate of 200 cubic inches per minute. How long will it take her to inflate the balloon? Round to the nearest tenth.

(Show your work here)

🌐 **Go Online** You can complete an Extra Example online.

Think About It!

How would you begin solving the problem?

_____

_____

_____

Talk About It!

Without calculating, how do you know that the time to inflate the volleyball will be greater than one minute?

Learn Volume of Hemispheres

A circle through the center of a sphere separates a sphere into two congruent halves each called a **hemisphere**.

Words	Model
The volume V of a hemisphere is two-thirds the product of π and the cube of the radius r.	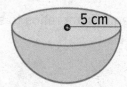

Symbols	
$V = \frac{2}{3}\pi r^3$	

Talk About It!

How are the coefficients $\frac{4}{3}$ and $\frac{2}{3}$ in the formulas for finding the volume of a sphere and a hemisphere related?

Example 4 Find Volume of Hemispheres

Find the volume of the hemisphere. Round to the nearest tenth.

5 cm

$V = \frac{2}{3}\pi r^3$ Volume of a hemisphere

$V = \frac{2}{3}\pi \left(\right)^3$ Replace r.

$V = \boxed{}\,\pi$ Multiply.

$V \approx \boxed{}$ Use a calculator.

So, the volume of the hemisphere is about 261.8 cubic centimeters.

Check

Find the volume of a hemisphere. Round to the nearest tenth.

16 in.

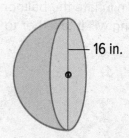

Go Online You can complete an Extra Example online.

🌐 Apply Packaging

Brad is packing 3 bouncy balls in a cylindrical container. The radius of each bouncy ball is 10 centimeters. The cylinder has a base area of 315 square centimeters and a height of 65 centimeters. What is the volume of empty space in the container? Round to the nearest tenth.

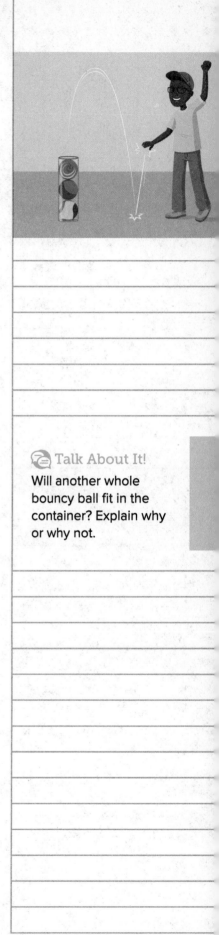

1 What is the task?

Make sure you understand exactly what question to answer or problem to solve. You may want to read the problem three times. Discuss these questions with a partner.

First Time Describe the context of the problem, in your own words.
Second Time What mathematics do you see in the problem?
Third Time What are you wondering about?

2 How can you approach the task? What strategies can you use?

3 What is your solution?

Use your strategy to solve the problem.

💬 Talk About It!

Will another whole bouncy ball fit in the container? Explain why or why not.

4 How can you show your solution is reasonable?

✍ Write About It! Write an argument that can be used to defend your solution.

Check

The radius of a table tennis ball is 2 centimeters. Olivia is packing 30 table tennis balls in a box with length of 24 centimeters, a width of 20 centimeters, and a height of 4 centimeters. What is the volume of the empty space? Round to the nearest tenth.

Ⓐ 914.7 cubic centimeters

Ⓑ 1,003.1 cubic centimeters

Ⓒ 1,005.3 cubic centimeters

Ⓓ 1,920 cubic centimeters

Show your work here

📍 **Go Online** You can complete an Extra Example online.

📖 **Foldables** It's time to update your Foldable, located in the Module Review, based on what you learned in this lesson. If you haven't already assembled your Foldable, you can find the instructions on page FL1.

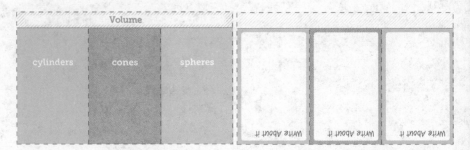

Practice

🡒 **Go Online** You can complete your homework online.

Find the volume of each sphere. Express your answer in terms of π. (Example 1)

1.

23 ft

2.
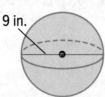
9 in.

3. A necklace has a single spherical pearl with a radius of 2.1 millimeters. What is the volume of the pearl? Round to the nearest tenth. (Example 2)

4. The radius of a mini-basketball is 4 inches. A pump can inflate the ball at a rate of 6 cubic inches per second. How long will it take to inflate the ball? Round to the nearest tenth. (Example 3)

Find the volume of each hemisphere. Round to the nearest tenth. (Example 4)

5.
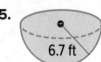
6.7 ft

6.
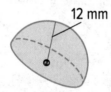
12 mm

Test Practice

7. Olga is using spherical beads to create a border on a picture frame. Each bead has a diameter of 1.5 millimeters. Find the volume of each bead. Round to the nearest tenth.

8. **Open Response** What is the volume of the sphere shown? (Use 3.14 for π.)

48 mm

9. Miguel has a ball of modeling clay that has a diameter of 3.5 centimeters. The cylindrical container it is placed in has a base area of 7.1 square centimeters and a height of 5 centimeters. What is the volume of empty space in the container? Round to the nearest tenth.

10. A gift set of three golf balls is packaged in a clear rectangular box 13.1 centimeters long, 4.5 centimeters wide, and 4.5 centimeters tall. If each ball is 4.3 centimeters in diameter, find the volume of the empty space in the box. Round to the nearest tenth.

11. MP Identify Structure When using a calculator to find the volume of a sphere, one way to calculate $\frac{4}{3}$ is to use the parentheses keys and multiply by $\left(\frac{4}{3}\right)$. What is another way to calculate $\frac{4}{3}$ when finding the volume of a sphere?

12. MP Persevere with Problems The circumference of a sphere is 18π inches. Find the volume of the sphere in terms of π.

13. Luci and Stefan are finding the volume of the hemisphere shown. Luci determines the volume to be 26.203π cubic feet and Stefan determines the volume to be 82.318 cubic feet. Whose answer is closer to the exact volume? Write an argument that can be used to defend your solution.

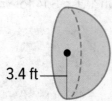

3.4 ft

14. MP Find the Error A student found the volume of the sphere shown. Find her mistake and correct it.

$\frac{4}{3}\pi(7.5)^3 \approx 1,767.1 \text{ in}^3$

7.5 in.

Find Missing Dimensions

I Can... use volume formulas to solve for missing dimensions in cones, cylinders, and spheres.

Learn Find Missing Dimensions Given the Volume

You can find missing dimensions of cylinders, cones, and spheres, given the volume and the other dimension(s) of the figure. Use the appropriate volume formulas and solve for the unknown value.

Volume Formulas		
Cylinder	Cone	Sphere
$V = Bh$ or $V = \pi r^2 h$	$V = \frac{1}{3}Bh$ or $V = \frac{1}{3}\pi r^2 h$	$V = \frac{4}{3}\pi r^3$

🔘 **Go Online** Watch the animation to see how to solve for a missing dimension when the volume is known.

The animation shows how to find the height of the cylinder shown, given the volume and radius.

$$V = \pi r^2 h \quad \text{Volume of a cylinder}$$
$$45\pi = \pi \cdot 3^2 \cdot h \quad V = 45\pi, r = 3$$
$$45\pi = \pi \cdot 9 \cdot h \quad \text{Simplify.}$$
$$45\pi = 9\pi h \quad \text{Multiply.}$$
$$\frac{45\pi}{9\pi} = \frac{9\pi h}{9\pi} \quad \text{Division Property of Equality}$$
$$5 = h \quad \text{Simplify.}$$

3 m

h

$V = 45\pi \text{ m}^3$

So, the height of the cylinder is 5 meters.

Example 1 Find Missing Dimensions of Cylinders

Think About It!

Which volume formula should you use?

The volume of a cylinder with a radius of 4 centimeters is 128π cubic centimeters.

What is the height of the cylinder?

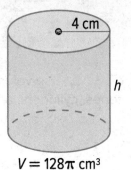

$V = 128\pi$ cm³

The volume and the radius are known.
You need to find the height.

Talk About It!

How does the volume of the cylinder change if its height is tripled? its radius?

$$V = \pi r^2 h$$ Volume of a cylinder

$$\boxed{} = \pi \left(\boxed{} \right)^2 h$$ Replace V and r.

$$128\pi = \boxed{}\pi h$$ Multiply.

$$\frac{128\pi}{16\pi} = \frac{16\pi h}{16\pi}$$ Division Property of Equality

$$\boxed{} = h$$ Simplify.

So, the height of the cylinder is 8 centimeters.

Check

The volume of a cylinder with a height of 12 centimeters is 300π cubic centimeters. What is the radius of the cylinder?

Show your work here

Go Online You can complete an Extra Example online.

Example 2 Find Missing Dimensions of Cones

The volume of a cone with a height of 6 inches is 8π cubic inches.

What is the radius of the cone?

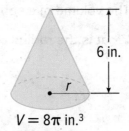

6 in.

$V = 8\pi$ in.³

Think About It!

Which volume formula should you use?

The volume and the height are given. You need to find the radius.

$V = \frac{1}{3}\pi r^2 h$ Volume of a cone

$\boxed{} = \frac{1}{3}\pi r^2 \left(\boxed{}\right)$ Replace V and h.

$8\pi = \boxed{}\,\pi r^2$ Multiply. $\frac{1}{3} \cdot 6 = 2$

$\dfrac{8\pi}{2\pi} = \dfrac{2\pi r^2}{2\pi}$ Division Property of Equality

$\boxed{} = r^2$ Simplify.

$\sqrt{4} = \sqrt{r^2}$ Take the positive square root of each side.

$\boxed{} = r$ Simplify.

So, the radius of the cone is 2 inches.

Talk About It!

How can you verify that the radius, 2 inches, is correct?

Check

The volume of a cone with a radius of 7.5 inches is 375π cubic inches. What is the height of the cone?

Show your work here

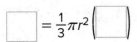 **Go Online** You can complete an Extra Example online.

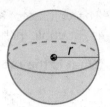

Example 3 Find Missing Dimensions of Spheres

The volume of a sphere is 972π cubic millimeters.

What is the radius of the sphere?

$V = 972\pi$ mm³

Suppose the volume of
a sphere is written as a
decimal, such as
3,053.63 cubic
millimeters. Is it
possible to find the
exact radius of the
sphere? Explain your
reasoning.

Use the volume formula for a sphere to solve for the radius.

$$V = \frac{4}{3}\pi r^3$$ Volume of a sphere

$$\boxed{} = \frac{4}{3}\pi r^3$$ Replace V.

$$\frac{972\pi}{\frac{4}{3}\pi} = \frac{\frac{4}{3}\pi r^3}{\frac{4}{3}\pi}$$ Division Property of Equality

$$\boxed{} = r^3$$ Simplify.

$$\sqrt[3]{729} = \sqrt[3]{r^3}$$ Take the cube root of each side.

$$\boxed{} = r$$ Simplify.

So, the radius of the sphere is 9 millimeters.

Check

The volume of a sphere is $\dfrac{62,500}{3}\pi$ cubic centimeters. What is the
radius of the sphere?

Show
your work
here

Go Online You can complete an Extra Example online.

🌐 Apply Shopping

A local grocery store sells corn in two different-sized cans. A one-meter wide shelf is being stocked. How many more of the smaller cans will fit on the shelf than the larger cans?

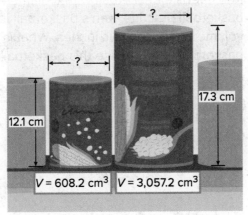

1 What is the task?

Make sure you understand exactly what question to answer or problem to solve.

You may want to read the problem three times. Discuss these questions with a partner.

First Time Describe the context of the problem, in your own words.
Second Time What mathematics do you see in the problem?
Third Time What are you wondering about?

2 How can you approach the task? What strategies can you use?

3 What is your solution?

Use your strategy to solve the problem.

💬 Talk About It!

Why is it necessary to round the number of larger cans to 6, rather than to 7?

4 How can you show your solution is reasonable?

✏️ **Write About It!** Write an argument that can be used to defend your solution.

Check

Ken needs a basketball with a diameter of at least 9 inches for basketball practice. Ken's basketball is labeled on the side with a volume of 448.9 cubic inches. What is the diameter of Ken's basketball? Can he use this basketball for practice?

(A) 8.75 inches; no

(B) 8.9 inches; no

(C) 9.5 inches; yes

(D) 11.97 inches; yes

> Show your work here

🅡 **Go Online** You can complete an Extra Example online.

Pause and Reflect

How does what you already know about solving equations help you with finding missing dimensions of cylinders, cones, and spheres?

> Record your observations here

Practice

Go Online You can complete your homework online.

1. The volume of a cylinder is 72π cubic feet and the radius is 6 feet. What is the height of the cylinder? (Example 1)

2. The volume of a cylinder is $5,070\pi$ cubic centimeters. The height of the cylinder is 30 centimeters. Find the radius. (Example 1)

3. The volume of a cone is 196π cubic feet. Its radius is 7 feet. Find the height. (Example 2)

4. The volume of a cone is 735π cubic millimeters and the height is 5 millimeters. What is the radius of the cone? (Example 2)

5. Find the radius of a sphere with a volume of $26,244\pi$ cubic inches. (Example 3)

6. The volume of a sphere is $4,500\pi$ cubic yards. What is the radius of the sphere? (Example 3)

Test Practice

7. Melody has a mug with a diameter of 3.5 inches and a height of 4 inches. It is filled to the top with water. She wants to pour it into a different mug with a diameter of 3 inches. What is the minimum height the different mug must be so it does not overflow? Round to the nearest tenth.

8. **Equation Editor** The volume of a sphere is $\frac{1,372}{3}\pi$ cubic inches. Find the diameter of the sphere, in inches.

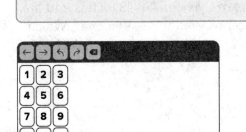

9. A party store is placing party hats in the shape of cones on a 6-foot long shelf. If the party hats are lined up in a row, how many more of the smaller party hats will fit on the shelf than the larger party hats?

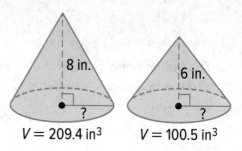

8 in.

6 in.

$V = 209.4$ in³ $V = 100.5$ in³

10. Consuela is displaying a cylindrical case on a pedestal. She wants it to be placed so that there is an even amount of space on either side of the case. If the volume of the cylinder shown is 1,130.97 cubic inches, how much space will be on either side if it's placed on a pedestal that is 18 inches across?

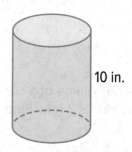

10 in.

11. 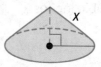 **Find the Error** A student was finding the radius of a sphere with a volume of $4,500\pi$ cubic inches. Find his mistakes and correct them.

$$V = \frac{4}{3}\pi r^3$$

$$4,500\pi = \frac{4}{3}\pi r^3$$

$$4,500 = \frac{4}{3}r^3$$

$$6,000 = r^3$$

$$r = 2,000$$

12. **Persevere with Problems** The volume of the cone shown is 240π cubic meters. The height of the cone is 5 meters. Find the length of the slant height, x.

x

13. **Be Precise** Two cylinders have the same volume of 845π cubic inches. The radius of Cylinder A is 13 inches and the radius of Cylinder B is 10 inches. Which cylinder is taller? How much taller?

14. When finding the volume of prisms and pyramids, a unit cube can be used. Why is a unit sphere not a good unit of volume measurement for all spheres?

Volume of Composite Solids

I Can... find the volume of a composite figure by decomposing it into cubes, cones, cylinders, and spheres, and using the known volume formulas for these figures.

What Vocabulary Will You Learn?
composite solids

Learn Composite Solids

Composite solids are objects that are composed of multiple three-dimensional solids.

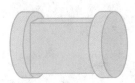

2 cylinders and a rectangular prism

2 cones and a cylinder

hemisphere and cone

sphere and cylinder

Learn Volume of Composite Solids

To find the volume of composite solids, decompose the object into solids whose volumes you know how to find.

Go Online Watch the animation to see how to find the volume of the composite solid shown.

Cone

$V = \frac{1}{3}\pi r^2 h$

$V = \frac{1}{3} \cdot \pi \cdot 12^2 \cdot 14$

$V = \frac{1}{3} \cdot \pi \cdot 144 \cdot 14$

$V = 672\pi$

Cylinder

$V = \pi r^2 h$

$V = \pi \cdot 12^2 \cdot 8$

$V = \pi \cdot 144 \cdot 8$

$V = 1{,}152\pi$

14 cm

12 cm

8 cm

So, the volume of the solid is $672\pi + 1{,}152\pi$ or about 5,730.3 cubic centimeters.

Think About It!

How would you begin solving the problem?

 Talk About It!

What is the benefit of keeping the volumes in terms of π until the final step when finding the volume of a composite solid?

Example 1 Find Volume of Composite Solids

Find the volume of the solid. Round to the nearest tenth.

Step 1 Find the volume of each solid.

Cylinder

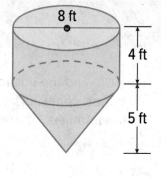

$V = \pi r^2 h$ Volume of a cylinder

$V = \pi(4)^2(4)$ Replace r and h.

$V = 64\pi$ Multiply.

The volume of the cylinder is _____ cubic feet.

Cone

$V = \frac{1}{3}\pi r^2 h$ Volume of a cone

$V = \frac{1}{3}\pi(4)^2(5)$ Replace r and h.

$V = \frac{80}{3}\pi$ Multiply.

The volume of the cone is _____ cubic feet.

Step 2 Add the volumes and simplify.

$64\pi + \frac{80}{3}\pi = \frac{272}{3}\pi$

$\frac{272}{3}\pi$ simplifies to _____ when it is rounded to the nearest tenth. So, the volume of the composite solid is about 284.8 cubic feet.

Check

A platform like the one shown was built to hold a sculpture for an art exhibit. What is the volume of the solid? Round to the nearest whole cubic meter.

Show your work here

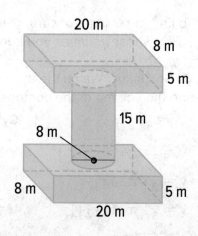

Go Online You can complete an Extra Example online.

🌐 Example 2 Find Volume of Composite Solids

Tanya uses cube-shaped beads to make jewelry. Each bead has a circular hole through the middle.

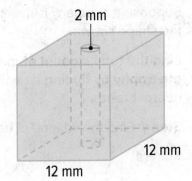

Find the volume of each bead. Round to the nearest tenth.

Step 1 Find the volume of each solid.

Cube

$V = s^3$	Volume of a cube
$V = (12)^3$	Replace s.
$V = 1{,}728$	Multiply.

The volume of the cube is _____ cubic millimeters.

Cylinder

$V = \pi r^2 h$	Volume of a cylinder
$V = \pi(1)^2(12)$	Replace r and h.
$V = 12\pi$	Multiply.

The volume of the cylinder is _____ cubic millimeters.

Step 2 Subtract to find the volume of the bead.

$1{,}728 - 12\pi \approx$ _____

So, the volume of the bead is about 1,690.3 cubic millimeters.

Check

Find the volume of the solid. Round to the nearest tenth.

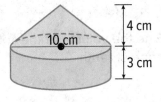

🅝 **Go Online** You can complete an Extra Example online.

💭 **Think About It!**

What three-dimensional objects make up the figure?

💬 **Talk About It!**

Suppose that Tanya decides to use a spherical bead with a diameter equal to the cube's side length. Is the volume of spherical bead greater or lesser than the volume of the cube-shaped bead?

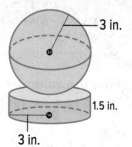

☁️ Think About It!

What dimensions are needed to find the volume of the object?

Suppose a company is making a solid trophy in the shape shown.

Find the total amount of metal used to make the trophy by finding its volume. Round to the nearest tenth.

Step 1 Find the volume of each solid.

Sphere

$$V = \frac{4}{3}\pi r^3 \qquad \text{Volume of a sphere}$$
$$V = \frac{4}{3}\pi(3)^3 \qquad \text{Replace } r.$$
$$V = 36\pi \qquad \text{Multiply.}$$

The volume of the sphere is _____ cubic inches.

Cylinder

$$V = \pi r^2 h \qquad \text{Volume of a cylinder}$$
$$V = \pi(3)^2(1.5) \qquad \text{Replace } r \text{ and } h.$$
$$V = 13.5\pi \qquad \text{Multiply.}$$

The volume of the cylinder is _____ cubic inches.

Step 2 Add the volumes and simplify.

$$36\pi + 13.5\pi = \boxed{}\pi$$

49.5π simplifies to _____ when it is rounded to the nearest tenth. So, the volume of the composite solid is about 155.5 cubic inches.

☁️ Talk About It!

Why is it best to express the volume rounded to the nearest tenth, in this situation, rather than in terms of π?

Check

Find the volume of the solid. Round to the nearest tenth.

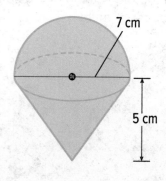

🔵 **Go Online** You can complete an Extra Example online.

🌎 Apply Art

Trevor is creating a concrete sculpture of an ice cream cone as shown. Trevor needs to order a whole number of cubic yards of concrete. How many cubic yards of concrete should Trevor order so that the amount left over is minimized?

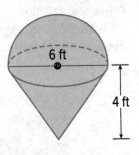

6 ft

4 ft

1 What is the task?

Make sure you understand exactly what question to answer or problem to solve. You may want to read the problem three times. Discuss these questions with a partner.

First Time Describe the context of the problem, in your own words.
Second Time What mathematics do you see in the problem?
Third Time What are you wondering about?

2 How can you approach the task? What strategies can you use?

Record your observations here

3 What is your solution?

Use your strategy to solve the problem.

Show your work here

4 How can you show your solution is reasonable?

✏️ **Write About It!** Write an argument that can be used to defend your solution.

💬 **Talk About It!**

How could this problem be solved another way?

Check

A jeweler is thinking about using this solid silver bead in a necklace he is designing. He only has 5,000 grams of silver. If pure silver's mass is 10.5 grams per cubic centimeter, can he make this bead? If so, how many grams of silver will the jeweler have left?

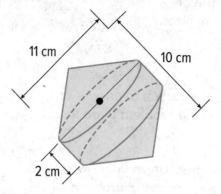

Show your work here

(A) no

(B) yes; 343.4 g

(C) yes; 1,267 g

(D) yes; 1,673.9 g

🔘 **Go Online** You can complete an Extra Example online.

Pause and Reflect

Did you have difficulty finding the volume of composite solids? If so, what can you do to get help? If not, how could you explain the process to another student?

Record your observations here

Practice

▶ **Go Online** You can complete your homework online.

Find the volume of each solid. Round to the nearest tenth. (Example 1)

1.

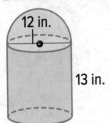

12 in.
13 in.

2.

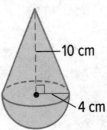

10 cm
4 cm

3. Find the volume of the flower vase. Round to the nearest tenth. (Example 2)

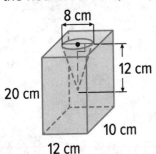

8 cm
12 cm
20 cm
10 cm
12 cm

4. Find the volume of the nail polish bottle. Round to the nearest tenth. (Example 3)

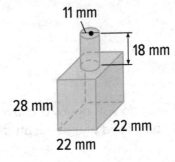

11 mm
18 mm
28 mm
22 mm
22 mm

5. Find the volume of the salt shaker. Round to the nearest tenth.

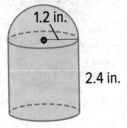

1.2 in.
2.4 in.

6. Find the volume of the solid. Round to the nearest tenth.

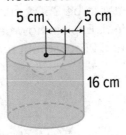

5 cm   5 cm
16 cm

7. Open Response A box contains six identical cans, as shown. What percentage of the volume of the box is occupied by the cans? Round to the nearest tenth of a percent.

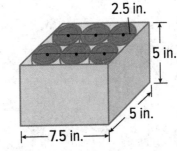

2.5 in.
5 in.
5 in.
7.5 in.

8. What is the volume of the composite solid in cubic yards? Round to the nearest tenth.

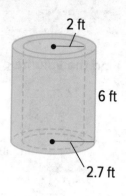

2 ft

6 ft

2.7 ft

9. Mel's Ice Cream Shop has cylindrical containers to hold its ice cream. Each cylinder has a diameter of 10 inches and a height of 15 inches. How many of the cones shown can be made without any leftover ice cream?

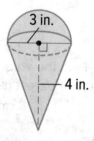

3 in.

4 in.

10. What measurements do you need to know in order to find the volume of a composite solid composed of a hemisphere and a cone?

11. (MP) **Be Precise** Mateo is finding the volume of the solid shown. He found the volume of the cylinder to be 250π cubic feet and the volume of the cone to be 25π cubic feet. Explain how he can use the Distributive Property to add 250π and 25π without using an approximation.

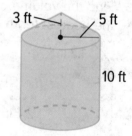

3 ft   5 ft

10 ft

12. (MP) **Find the Error** A student found the volume of the solid shown. Find her mistake and correct it.

$$V = \frac{4}{3}\pi r^3 + \pi r^2 h$$

$$V = \frac{4}{3}\pi(3)^3 + \pi(3)^2(15)$$

$$V = 171\pi \text{ yd}^3$$

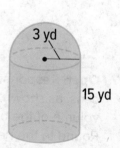

3 yd

15 yd

Foldables Use your Foldable to help review the module.

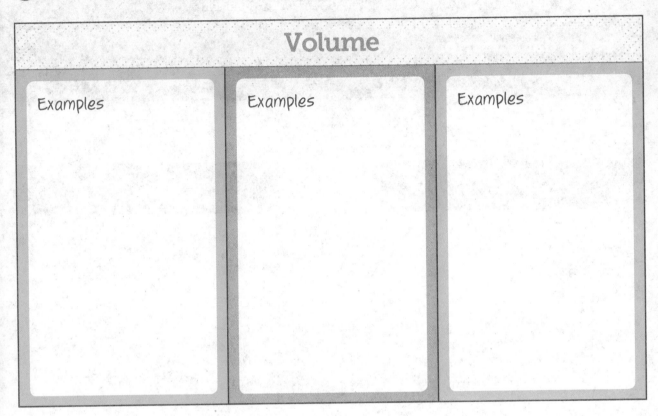

Rate Yourself! ⬤ ◈ ★

Complete the chart at the beginning of the module by placing a checkmark in each row that corresponds with how much you know about each topic after completing this module.

Write about one thing you learned.

_____

_____

_____

_____

_____

Write about a question you still have.

_____

_____

_____

_____

_____

Reflect on the Module

Use what you learned about volume to complete the graphic organizer.

e Essential Question

How can you measure a cylinder, cone, or sphere?

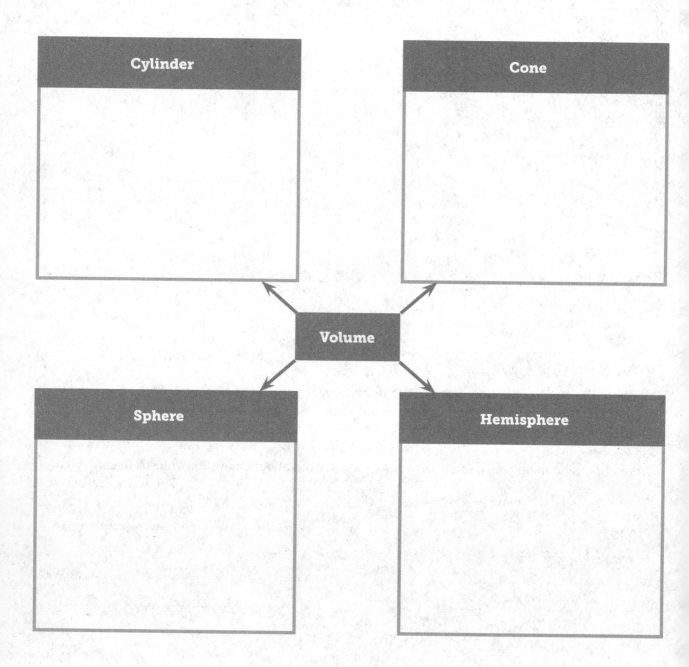

Cylinder

Cone

Volume

Sphere

Hemisphere

Test Practice

1. Equation Editor Find the volume of the cylinder in cubic inches. Round to the nearest tenth. (Lesson 1)

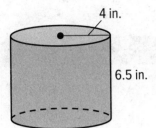

4 in.

6.5 in.

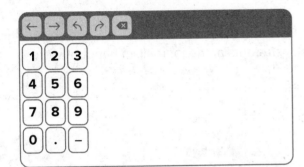

2. Multiple Choice A galvanized stock tank with the dimensions shown is filling with water at a rate of 25 gallons per minute. About how many minutes will it take to fill the stock tank if 1 cubic foot is about 7.5 gallons? Round to the nearest minute.

(Lesson 1)

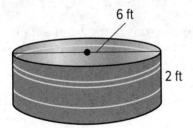

6 ft

2 ft

- Ⓐ 10 minutes
- Ⓑ 17 minutes
- Ⓒ 34 minutes
- Ⓓ 68 minutes

3. Open Response Find the volume of the cone. Express your answer in terms of π.

(Lesson 2)

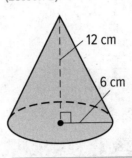

12 cm

6 cm

4. Open Response A conical and a cylindrical fish tank are shown. (Lesson 2)

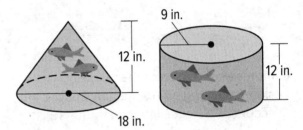

9 in.

12 in.

12 in.

18 in.

A. Suppose both aquariums are filled with water. Find the volume of each aquarium. Round each calculation to the nearest whole cubic inch.

B. If 1 cubic inch of water weighs 0.6 ounce, about how many more ounces does the water in the cylindrical aquarium weigh?

- Ⓐ 1,014 ounces
- Ⓑ 1,221.6 ounces
- Ⓒ 2,036 ounces
- Ⓓ 2,443 ounces

5. Equation Editor Find the volume of the sphere in cubic inches. Round to the nearest tenth. (Lesson 3)

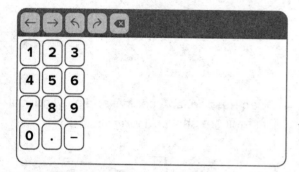

6. Multiselect Which of the following statements regarding the hemisphere are accurate? Select all that apply. (Lesson 3)

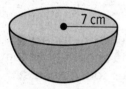

☐ The diameter of the hemisphere is 14 centimeters.

☐ The volume of a sphere is half the volume of a hemisphere that has the same radius.

☐ The volume of a hemisphere is half the volume of a sphere that has the same radius.

☐ The volume of the hemisphere, rounded to the nearest tenth, is 718.4 cubic centimeters.

☐ The volume of the hemisphere, rounded to the nearest tenth, is 205.3 cubic centimeters.

7. Equation Editor The volume of a cylinder with a radius of 8 feet is 192π. What is the height of the cylinder in feet? (Lesson 4)

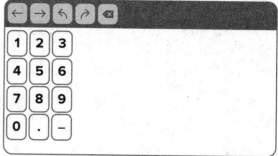

8. Open Response The awards received by the students at the journalism banquet were shaped as shown. (Lesson 5)

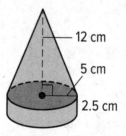

A. Find the total volume of the award. Round to the nearest tenth.

B. Suppose the award is made from a high density polyethylene that has a density of 0.95 gram per cubic centimeter. What is the mass of the award? Round to the nearest whole number.

Ⓐ 265 grams

Ⓑ 342 grams

Ⓒ 430 grams

Ⓓ 485 grams

Scatter Plots and Two-Way Tables

e Essential Question

What do patterns in data mean and how are they used?

What Will You Learn?

Place a checkmark (✓) in each row that corresponds with how much you already know about each topic **before** starting this module.

KEY	Before			After		
⬛ — I don't know. ◆ — I've heard of it. ★ — I know it!	⬛	◆	★	⬛	◆	★
constructing scatter plots						
interpreting scatter plots						
drawing lines of fit						
making conjectures using lines of fit						
writing equations for lines of fit						
constructing two-way tables						
finding and interpreting relative frequencies						
finding relative frequencies to determine associations						

📦 Foldables Cut out the Foldable and tape it to the Module Review at the end of the module. You can use the Foldable throughout the module as you learn about scatter plots and two-way tables.

What Vocabulary Will You Learn?

Check the box next to each vocabulary term that you may already know.

☐ bivariate data   ☐ relative frequency

☐ cluster   ☐ scatter plot

☐ line of fit   ☐ two-way table

☐ outlier

Are You Ready?

Study the Quick Review to see if you are ready to start this module.
Then complete the Quick Check.

Quick Review

Example 1
Identify slopes and y-intercepts from graphs.

Find the slope and y-intercept of the line.

The rise is 3 and the run is 1. The slope is $\frac{\text{rise}}{\text{run}}$ or $\frac{3}{1}$ or 3.

The line crosses the y-axis at (0, −1), so the y-intercept is −1.

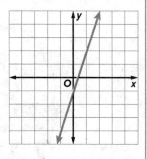

Example 2
Interpret data in tables.

The table shows the survey results of students' favorite colors. How many students were surveyed?

Color	Red	Blue	Green
Students	34	45	16

34 + 45 + 16 = 95   Add the number of students that voted for each color.

So, 95 students were surveyed.

Quick Check

1. Find the slope and y-intercept of the line.

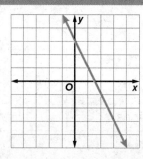

2. Use the table from Example 2. How many more students voted for blue as their favorite color compared to green?

How Did You Do?
Which exercises did you answer correctly in the Quick Check?
Shade those exercise numbers at the right.

Scatter Plots

I Can... use a set of bivariate data to construct a scatter plot and describe the association as positive or negative and as linear or nonlinear.

What Vocabulary Will You Learn?
bivariate data

cluster

outlier

scatter plot

Explore Scatter Plots

Online Activity You will explore how to construct and interpret scatter plots.

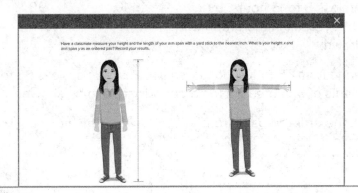

Have a classmate measure your height and the length of your arm span with a yard stick to the nearest inch. What is your height x and arm span y as an ordered pair? Record your results.

Learn Construct Scatter Plots

Bivariate data consists of two variables or two numerical observations. A **scatter plot** is a graph that shows the relationship between bivariate data. The bivariate data is graphed as ordered pairs on the coordinate plane.

Go Online Watch the animation to see how to construct a scatter plot to display the following data.

Time (h)	2.5	3	4	4.25	5.5	6	7
Distance (mi)	5	6.5	6.75	7	8.75	9	9.5

Let x represent the number of hours and y represent the distance hiked. Graph the ordered pairs from the table.

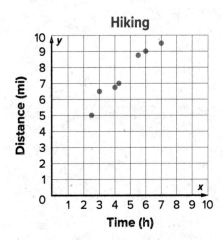

Hiking

💬 Talk About It!

Is there a relationship between the number of hours spent practicing and the test score? How do you know?

🌐 **Example 1** Construct Scatter Plots

Construct a scatter plot of the number of hours students spent practicing for a driving test and their score on the driver's test.

Number of Hours	Test Score (points)
6	65
7	70
12	100
8	80
9	87
11	93
10	90

Step 1 Determine appropriate labels and a reasonable scale for the axes.

Let the x-axis represent the number of hours, and let the y-axis represent the test score.

The hours range from 6 to 12, so a reasonable scale is 5 to 13 hours, by increments of 1. A break in the graph will need to be shown.

The scores range from 65 to 100 points, so a reasonable scale is 65 to 100 by increments of 5. A break in the graph will need to be shown.

Step 2 Graph the ordered pairs from the table.

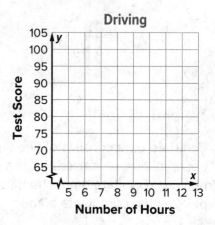

Driving

Check

Construct a scatter plot of the number of millions of viewers who watched new seasons of a certain television show.

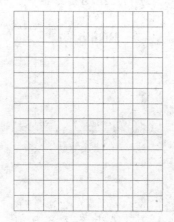

Season	Viewers (millions)
1	31.7
2	26.3
3	25.0
4	24.7
5	22.6
6	22.1

Learn Interpret Scatter Plots

You can analyze the shape of the distribution of a scatter plot to investigate patterns of association between the variables.

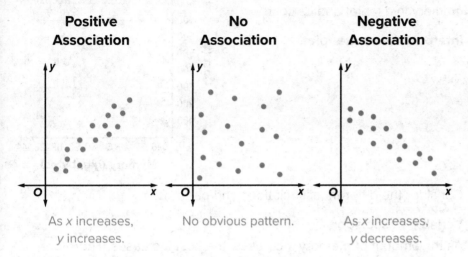

Positive Association	**No Association**	**Negative Association**
As x increases, y increases.	No obvious pattern.	As x increases, y decreases.

If the distribution of points on a scatter plot shows a positive or negative association, then the distribution can be classified as linear or nonlinear.

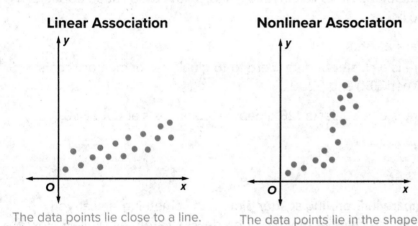

Linear Association

The data points lie close to a line.

Nonlinear Association

The data points lie in the shape of a curve.

The scatter plot below shows a positive nonlinear association. A **cluster** is a collection of points that are close together. An **outlier** is a point that does not fit the pattern.

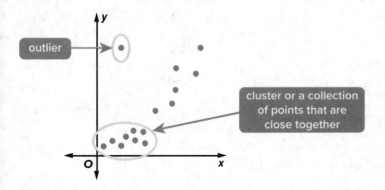

outlier

cluster or a collection of points that are close together

Talk About It!

What are some real-world examples where the points on a scatter plot would have a positive, linear relationship?

🌐 Example 2 Interpret Scatter Plots

The scatter plot shows the relationship between the amount of memory in a tablet and its cost.

Interpret the scatter plot.

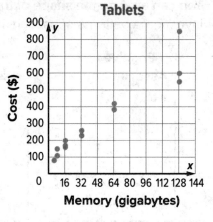

Tablets

Consider the different associations and patterns.

Variable Association

As the amount of memory increases, the cost increases. Therefore, the scatter plot shows a positive association.

Linear Association

The data appear to lie close to a line, so the association appears to be linear.

Other Patterns

There is a cluster of data. Zero to 16 gigabytes of memory costs between $50 and $200.

There is one outlier, a 128 gigabyte tablet costs about $850.

Pause and Reflect

Compare interpreting scatter plots with something similar you learned in an earlier module or grade. How are they similar? How are they different?

Record your observations here

Check

Which is the best interpretation of the scatter plot shown?

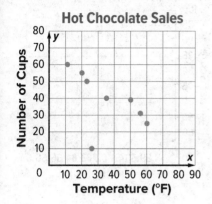

Hot Chocolate Sales

A) As temperature increases, the number of cups sold decreases, so the scatter plot shows a negative association. The association is linear. There are no clusters but there is one outlier. When the temperature was about 26°F, about 10 cups of hot chocolate were sold.

B) As temperature increases, the number of cups sold decreases, so the scatter plot shows a positive association. The association is linear. There are no clusters but there is one outlier. When the temperature was about 26°F, about 10 cups of hot chocolate were sold.

C) As temperature increases, the number of cups sold decreases, so the scatter plot shows a negative association. The association is nonlinear. There are no clusters but there is one outlier. When the temperature was about 26°F, about 10 cups of hot chocolate were sold.

D) As temperature increases, the number of cups sold decreases, so the scatter plot shows a negative association. The association is linear. There are no clusters and no outliers.

Math History Minute

18th century Spanish mathematician **María Andrea Casamayor (1700–1780)** wrote and published mathematics books which were then used in Spanish schools. Her first book, *Tirocinio Aritmético*, stated the basic rules of arithmetic and included tables of equivalent measures. Her second book, *El Para Solo*, included applications of mathematics in everyday life.

Go Online You can complete an Extra Example online.

🌐 Example 3 Interpret Scatter Plots

The scatter plot shows the relationship between a person's height and their most recent math test score.

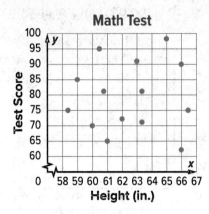

Math Test

Interpret the scatter plot.

Consider the different associations and patterns.

Variable Association

There does not appear to be any assocation between the variables.

Linear Association

Since the data do not lie close to a line or curve, there is no linear or nonlinear association.

Other Patterns

There are no clusters or outliers.

Check

Which is the best interpretation for the scatter plot shown?

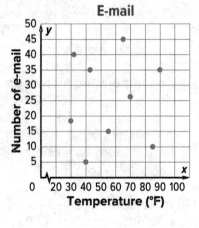

E-mail

Ⓐ The number of e-mail messages someone receives increases as the temperature increases. Therefore, there is a positive, linear association. There are no clusters or outliers.

Ⓑ The number of e-mail messages someone receives in a day does not depend on the temperature outside. Therefore, the scatter plot shows no association.

Ⓒ The number of e-mail messages someone receives decreases as the temperature increases. Therefore, there is a negative, linear association. There are no clusters or outliers.

Ⓓ There are no clusters or outliers. Therefore, there is a linear association.

🧭 **Go Online** You can complete an Extra Example online.

🌐 Apply Shopping

The table shows data that was collected from a random survey for the amount of time different people spent in a grocery store and the amount of money they spent. Interpret a scatter plot representing the data.

Time in Store (min)	15	20	30	40	35	25	20	35	20
Money Spent ($)	10	16	45	65	60	30	90	50	20

1 What is the task?

Make sure you understand exactly what question to answer or problem to solve. You may want to read the problem three times. Discuss these questions with a partner.

First Time Describe the context of the problem, in your own words.
Second Time What mathematics do you see in the problem?
Third Time What are you wondering about?

2 How can you approach the task? What strategies can you use?

Record your observations here

3 What is your solution?

Use your strategy to solve the problem.

Show your work here

💬 Talk About It!

Suppose a person was in the store for 30 minutes but left without buying anything. How would the scatter plot change?

4 How can you show your solution is reasonable?

✏️ **Write About It!** Write an argument that can be used to defend your solution.

Check

The table shows data that was collected from a random survey for the number of hours different people work in a week and their age.

Age (years)	16	20	32	35	18	22	26	17	25
Hours Worked	3	12	40	45	7	13	25	6	50

If a scatter plot was constructed to represent the data, which interpretation would best represent it?

Ⓐ As the age of a person increases, the number of hours they work in a week decreases. Therefore, the scatter plot shows a negative association. The association is linear. There is a cluster of data. People that are 16 to 22 years of age tend to work less than 15 hours. There is one outlier. Someone that is 25 years old works 50 hours a week.

Ⓑ As the age of a person increases, the number of hours they work in a week increases. Therefore, the scatter plot shows a positive association. The association is linear. There is a cluster of data. People that are 16 to 22 years of age tend to work less than 15 hours. There is one outlier. Someone that is 25 years old works 50 hours a week.

Ⓒ As the age of a person increases, the number of hours they work in a week increases. Therefore, the scatter plot shows a positive association. The association is linear. There are no clusters or outliers.

Ⓓ As the age of a person increases, the number of hours they work in a week decreases. Therefore, the association is nonlinear.

🐦 **Go Online** You can complete an Extra Example online.

📙 **Foldables** It's time to update your Foldable, located in the Module Review, based on what you learned in this lesson. If you haven't already assembled your Foldable, you can find the instructions on page FL1.

Practice

📍 **Go Online** You can complete your homework online.

1. The table shows the average points scored per game by an NBA player in the first ten seasons of his career. Construct a scatter plot of the data. (Example 1)

Season	1	2	3	4	5
Average Points Per Game	28.2	22.7	37.1	35.0	32.5
Season	6	7	8	9	10
Average Points Per Game	33.6	31.5	30.1	32.6	26.9

Test Practice

2. The scatter plot shows the relationship between the number of pieces in a jigsaw puzzle and the number of minutes that are recommended to complete the puzzle. Interpret the scatter plot. (Example 2)

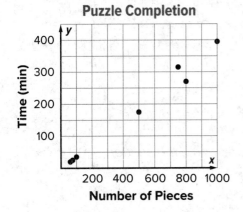

3. **Multiple Choice** The scatter plot shows the relationship between the birth month of every student in Mari's class and their height. Which is the best interpretation of the data? (Example 3)

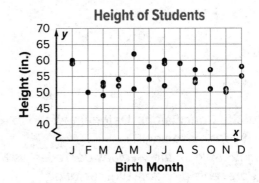

Ⓐ As the months progress, the heights of the students increase. There is a positive, linear association. There are no clusters or outliers.

Ⓑ The height of a student does not depend on their birth month. The scatter plot shows no association.

Ⓒ As the months progress, the heights of the students decrease. There is a negative, linear association. There are no clusters or outliers.

Ⓓ As the months progress, the heights of the students are the same. There is a positive, linear association.

Apply

4. The table shows the relationship between the number of days of school missed by students and their semester grades. Interpret a scatter plot representing the data.

Days Missed	8	3	2	10	6	7	1	13	11	4
Semester Grade	70	84	92	72	72	81	95	71	69	80
Days Missed	1	13	4	6	3	5	12	3	6	2
Semester Grade	98	68	91	72	91	78	70	89	76	94

5. **MP** **Find the Error** The table shows the daily high temperature and the number of cups of lemonade sold at a concession stand that day. Lucas determined that a scatter plot of the data would show that as the temperature increases, the number of cups sold decreases. Find his mistake and correct it.

Temperature (°F)	Cups Sold	Temperature (°F)	Cups Sold
80	12	98	40
72	7	77	18
89	26	67	5
93	37	82	19
74	7	86	16

6. **MP** **Justify Conclusions** Determine if the following statement is *true* or *false*. Explain your reasoning.

In a scatter plot, if the y-values decrease as the x-values decrease, the scatter plot represents a negative association.

7. **Create** Describe a situation that the scatter plot shown might represent. Then interpret the scatter plot.

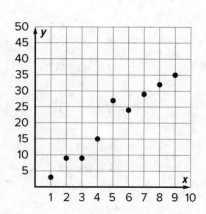

Draw Lines of Fit

I Can... use a scatter plot to draw a line that closely fits the data and predict values that are not present in the original data set.

What Vocabulary Will You Learn?
line of fit

Explore Lines of Fit

Online Activity You will use the Coordinate Graphing eTool to explore how to draw lines that fit a set of data.

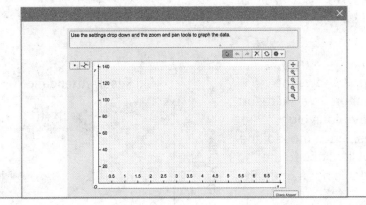

Learn Lines of Fit

When data are collected, the points graphed usually do not form a straight line, but may approximate a linear relationship. A **line of fit** is a line that goes through the center and is very close to most of the data points. It models the relationship between the two variables on the scatter plot.

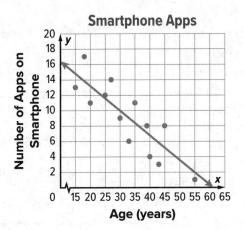

(continued on next page)

Go Online Watch the animation to see how to draw and assess a line of fit for each of the following data.

The scatter plot shows the weights and ages of baby elephants. To draw a line of fit, draw a straight line so that about half of the points are above the line and half of the points are below the line. Since most of the points lie close to the line, the model is a good fit.

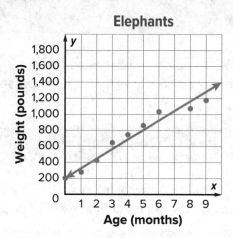

The scatter plot shows the number of visitors to a ski resort on days with various high temperatures. Since most of the points do not lie close to the line, the model is not a good fit.

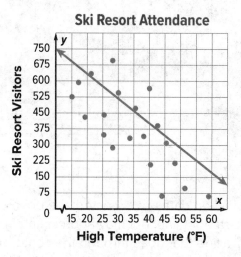

Talk About It!

How can you determine if the line shown in the graph on page 591 fits the data?

The scatter plot shows the relationship between math test scores and the number of hours certain students slept on the night before the test. When drawing a line of fit, ignore the outlier at point (4, 99). Other than the outlier, most of the points lie close to line. So, the model is a good fit.

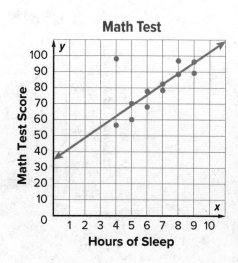

Example 1 Draw Lines of Fit

The table shows the number of hours people trained in a month and their 5K race time.

Draw and assess a line that seems to represent the data.

Hours Trained	5	8	12	18	25	30	32	23
Race Time (min)	41	34	35	27	28	21	16	20

Think About It!

Do you think the data in the table shows a positive, negative, or no association?

Part A Draw a line that fits most of the data.

Step 1 Construct a scatter plot.

Graph the ordered pairs from the table.

Step 2 Draw the line.

Does there appear to be a linear association between the variables?

_____

Draw a line that fits the data.

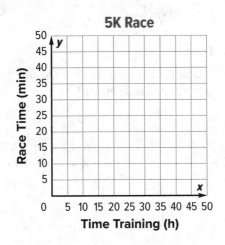

5K Race

Race Time (min) / Time Training (h)

Talk About It!

Is there only one correct answer when drawing a line that fits the data? Explain.

Part B Assess the line of fit.

Do most of the points lie close to the line? _____

Since most of the points lie close to the line, the model is a good fit.

Pause and Reflect

Did you have difficulty drawing a line of fit? If so, what can you do to get help? If not, how could you explain the process to another student?

Record your observations here

Check

The table shows the average traffic volume and average vehicle speed on a certain highway in a week.

Average Traffic Volume	Average Vehicle Speed (mph)
6,300	45
5,500	50
6,000	53
5,200	58
7,100	43
7,000	38
7,600	37

Part A

Construct a scatter plot on the graph provided. Use a ruler to approximate a line of fit.

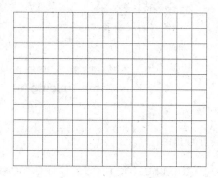

Part B

Assess the line of fit.

🎯 **Go Online** You can complete an Extra Example online.

Learn Make Conjectures Using Lines of Fit

A line of fit can be used to make conjectures about data that are not graphed on a scatter plot.

The scatter plot shows the distance traveled and the cost of an airline ticket. How can you use the line of fit to make a conjecture about the cost of an airline ticket if the distance traveled is 1,300 miles?

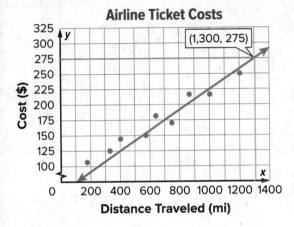

Airline Ticket Costs

If the distance traveled is 1,300 miles, then it is reasonable to expect that the cost of the airline ticket would be around $275.

🌐 Example 2 Make Conjectures Using Lines of Fit

The scatter plot shows the diameter of different trees and their heights.

Use the line of fit that is drawn to make a conjecture about the height of a tree if its diameter is 8 inches.

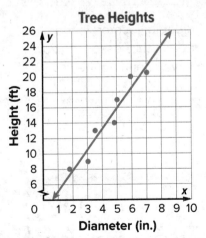

Tree Heights

Using the line, the *y*-value that corresponds with an *x*-value of 8 is about _____. So, you can predict that a tree that has a diameter of 8 inches will have a height of about 24 feet.

Check

The scatter plot shows the amount of money earned by different servers at a restaurant and the number of tables they served in a week. Use the line of fit that is drawn to make a conjecture about the amount of money a server will earn if they serve 225 tables in a week.

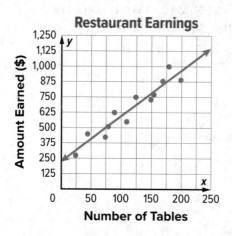

Restaurant Earnings

Go Online You can complete an Extra Example online.

Foldables It's time to update your Foldable, located in the Module Review, based on what you learned in this lesson. If you haven't already assembled your Foldable, you can find the instructions on page FL1.

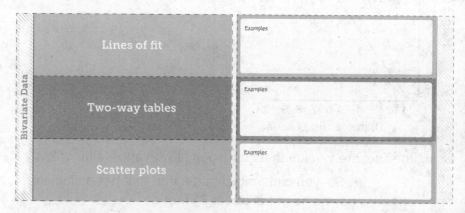

Practice

⊘ **Go Online** You can complete your homework online.

1. The table shows the average combined miles per gallon (MPG) and greenhouse gas (GHG) rating for certain mid-size cars. Construct a scatter plot. Then draw and assess a line that seems to represent the data. (Example 1)

Average MPG	22	25	31	28	16	26
GHG Rating	5	6	7	7	3	6
Average MPG	35	41	24	32	30	23
GHG Rating	8	9	5	8	7	5

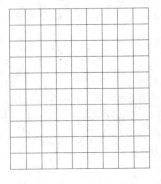

2. The table shows the fat and Calorie content for several snack foods. Construct a scatter plot. Then draw and assess a line that seems to represent the data. (Example 1)

Fat (g)	1	6	7	8	12	18	20
Calories	200	222	239	274	338	339	385

3. The scatter plot shows the number of cups of hot chocolate sold at a football game and the average temperature during the game. Use the line of fit to make a conjecture about the number of cups of hot chocolate sold if the average temperature is 50°F. (Example 2)

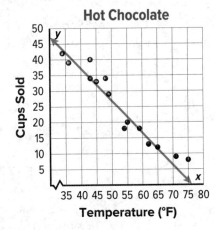

Hot Chocolate

4. The scatter plot shows the height and shoe size of the players on the boys' basketball team. Use the line of fit to make a conjecture about the shoe size of a boy on the team that is 59 inches tall. (Example 2)

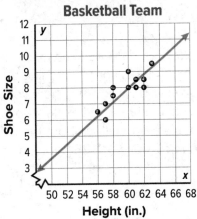

Basketball Team

5. **Open Response** The scatter plot shows the weight of different cars and their cost. Determine whether or not the line of fit is an accurate representation of the relationship between the data. Explain your reasoning.

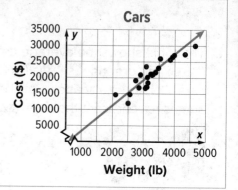

Cars

Apply

6. Several families are driving to an out-of-town soccer game. The table shows the distance each family drove and the time it took them. Construct a scatter plot to represent the data. Then draw and assess a line that seems to represent the data. Use the line of fit to make a conjecture about the time it would take a family to drive 135 miles to the game.

Distance (mi)	126	137	124	130	134	113
Time (h)	1.9	2.1	2.0	2.0	2.3	1.8
Distance (mi)	119	145	128	138	110	142
Time (h)	2.0	2.3	2.1	2.2	1.7	2.2

7. **MP** **Justify Conclusions** Determine if the following statement is *sometimes*, *always*, or *never* true. Explain your reasoning.
   A line of fit can be used to make a prediction about the data shown in a scatter plot.

8. The scatter plot shows the relationship between the number of hours different students spent watching television and their test score the following day. Describe how you would draw a line of fit for the relationship. Then explain how you can use that line of fit to make a prediction about the test score of a student that watched television for 2.25 hours the night before.

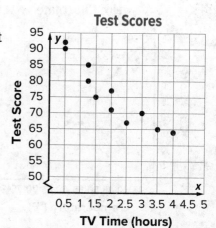

Test Scores

Equations for Lines of Fit

I Can... find the equation for a line that closely fits the data and use it to predict values that are not present in the original data set.

Learn Equations for Lines of Fit

Go Online Watch the animation to learn how to write an equation that approximates a set of data.

The animation shows the scatter plot of the amount of food collected and the number of volunteers who participated. You can write an equation in slope-intercept form for the line that is drawn.

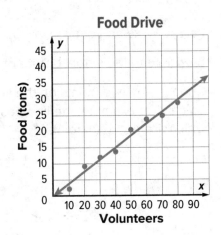

Food Drive

Step 1 Find the slope, or rate of change, of the line. Choose any two points on the line.

$m = \dfrac{y_2 - y_1}{x_2 - x_1}$ Definition of slope

$m = \dfrac{30 - 15}{80 - 40}$ $(x_1, y_1) = (40, 15)$ and $(x_2, y_2) = (80, 30)$

$m = \dfrac{15}{40}$ or 0.375   Simplify.

The slope is $\dfrac{15}{40}$ or 0.375. This means that there are 375 thousandths of a ton of food collected for every volunteer.

Step 2 Determine the y-intercept of the line drawn.

The y-intercept is 0 because the line crosses the y-axis at about the point (0, 0). This means that zero tons of food are collected when there are zero volunteers.

Step 3 Write the equation for the line drawn.

$y = mx + b$ Slope-intercept from

$y = 0.375x + 0$ Replace m with 0.375 and b with 0.

Talk About It!

Is it possible to have more than one equation for a line that approximates the data? Explain.

Think About It!

What can you say about the slope of the line drawn? What does that tell you about the relationship between time spent practicing and the number of mistakes made?

Example 1 Equations for Lines of Fit

The scatter plot shows the amount of time Mia spends practicing the piano and the number of mistakes made.

Write an equation in slope-intercept form for the line drawn that approximates the data. Then interpret the slope and *y*-intercept.

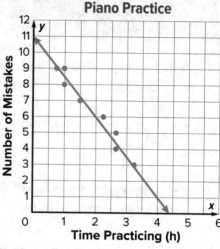

Piano Practice

Part A Write an equation in slope-intercept form for the line drawn.

Step 1 Find the slope, or rate of change, of the line.

Choose any two points on the line. They may or may not be data points. For example, the line passes through points (2, 6) and (4, 1), which are not data points.

$$m = \frac{y_2 - y_1}{x_2 - x_1}$$ Definition of slope

$$m = \frac{1 - 6}{4 - 2}$$ $(x_1, y_1) = (2, 6)$ and $(x_2, y_2) = (4, 1)$

$$m = \frac{\boxed{}}{\boxed{}} \text{ or } -2.5$$ Simplify.

Step 2 Determine the *y*-intercept and write the equation.

The *y*-intercept is about 11 because the line appears to cross the *y*-axis at the point (0, 11).

$y = mx + b$ Slope-intercept form

$y = \boxed{}x + \boxed{}$ Replace *m* with −2.5 and *b* with 11.

So, the equation for the line drawn is $y = -2.5x + 11$.

Part B Interpret the slope and *y*-intercept.

The slope is about −2.5. This means that Mia makes about _____ fewer mistakes each hour she practices. The *y*-intercept is about 11. This means Mia will make about _____ mistakes if she practices 0 hours.

Talk About It!

The equation for the line drawn is $y = -2.5x + 11$. Why is the slope of the line negative?

Check

The scatter plot shows the amount of time different students spend studying for a test and their test score.

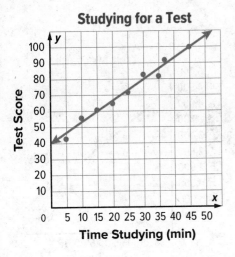

Studying for a Test

Part A

Which equation best represents the line that is drawn?

(A) $y = 0.66x + 40$

(B) $y = 1.33x - 40$

(C) $y = 1.33x + 40$

(D) $y = 0.66x - 40$

Part B

Interpret the slope and *y*-intercept.

 Go Online You can complete an Extra Example online.

Learn Make Conjectures Using Equations for Lines of Fit

You can use an equation that closely fits a set of data to make conjectures about the data.

The scatter plot shows the number of text messages that are sent and received in a month by different people and their age.

The equation $y = -80x + 4,500$ can be used to approximate the data.

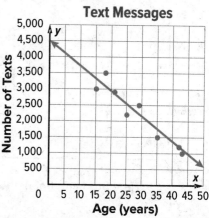

To predict how many text messages a person who is 55 years old will send and receive, substitute 55 for x into the equation and simplify.

$y = -80x + 4,500$ Write the equation.

$y = -80\left(\boxed{}\right) + 4,500$ Replace x with 55.

$y = \boxed{}$ Simplify.

So, it is reasonable to predict that a person who is 55 years old will send and receive about 100 text messages per month.

Pause and Reflect

Explain why you can use the equation for a line of fit to make conjectures about the data in a scatter plot.

Example 2 Make Conjectures Using Equations for Lines of Fit

The scatter plot shows the amount of money a company spends on advertising and the amount of money they make in sales over several months.

Write an equation for the line of fit drawn. Then use it to make a conjecture about the amount of money the company will make in sales if they spend $60,000 on advertising.

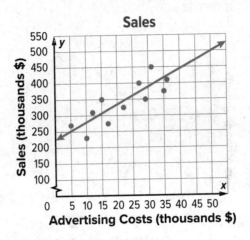

Sales

Sales (thousands $) vs Advertising Costs (thousands $)

😮 **Think About It!**

Will the sales be greater than $500,000? Justify your response.

Part A Write an equation for the line of fit drawn.

Step 1 Find the slope, or rate of change, of the line.

Choose any two points on the line. They may or may not be data points. The line passes through points (5, 250) and (50, 500).

$$m = \frac{y_2 - y_1}{x_2 - x_1} \qquad \text{Definition of slope}$$

$$m = \frac{500 - 250}{50 - 5} \qquad (x_1, y_1) = (5, 250) \text{ and } (x_2, y_2) = (50, 500)$$

$$m = \frac{250}{45} \text{ or about } 5.56 \qquad \text{Simplify.}$$

Step 2 Determine the y-intercept and write the equation.

The line of fit appears to cross the y-axis at the point (0, 225), so the y-intercept is 225.

$$y = mx + b \qquad \text{Slope-intercept form}$$

$$y = \boxed{}\, x + \boxed{} \qquad \text{Replace } m \text{ with 5.56 and } b \text{ with 225.}$$

So, the equation for the line of fit is $y = 5.56x + 225$.

Part B Use the equation for the line of fit to make a conjecture.

To make a conjecture about the amount of money made in sales if $60,000 is spent on advertising, replace x with 60.

$$y = 5.56x + 225 \qquad \text{Equation for line of fit}$$

$$y = 5.56\left(\boxed{}\right) + 225 \qquad \text{Replace } x \text{ with 60.}$$

$$y = \boxed{} \qquad \text{Simplify.}$$

So, if the company spends $60,000 on advertising, it will make about $558,600 in sales.

🗨 **Talk About It!**

In Part B, why is the answer $558,600, and not $558.60?

Check

The scatter plot shows the weight of different cars and their average gas mileage.

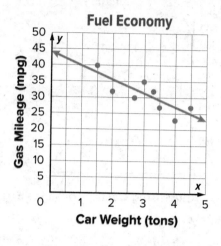

Fuel Economy

Part A

Which equation best represents the line of fit that is drawn?

(A) $y = -4.3x + 44$

(B) $y = -4.3x - 44$

(C) $y = -5x + 44$

(D) $y = -5x - 44$

Part B

Use the equation for the line of fit to make a conjecture about the gas mileage of a car if it weighs 5.5 tons.

🡒 **Go Online** You can complete an Extra Example online.

Apply Race Training

The scatter plot shows the number
of hours different people trained for
a 5K race and their time. How much
faster is the 5K race time expected
to be for a person who trains
35 hours, as opposed to 12 hours?

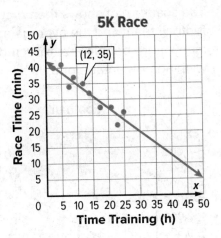

5K Race

(12, 35)

Race Time (min)

Time Training (h)

1 What is the task?

Make sure you understand exactly what question to answer or
problem to solve. You may want to read the problem three times.
Discuss these questions with a partner.

First Time Describe the context of the problem, in your own words.
Second Time What mathematics do you see in the problem?
Third Time What are you wondering about?

**2 How can you approach the task? What strategies can you
use?**

Record your
observations
here

3 What is your solution?

Use your strategy to solve the problem.

Show
your work
here

Talk About It!

How could you
solve the problem
another way?

4 How can you show your solution is reasonable?

 Write About It! Write an argument that can be used to defend
your solution.

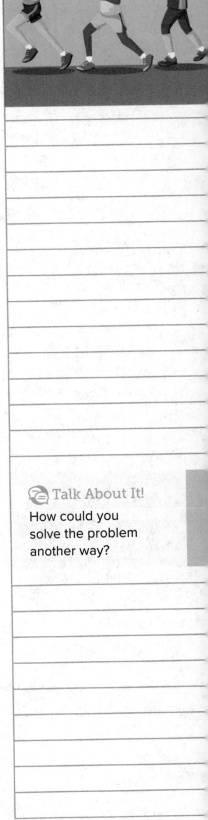

Check

The scatter plot shows the percent of games won by a high school football team and the average attendance for several seasons. How many more people are expected to attend games for a team with a winning percentage of 0.85 as opposed to 0.30?

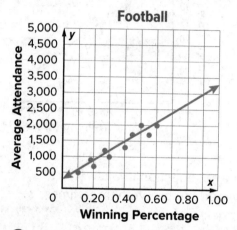

Football

A) 800

B) 900

C) 1,200

D) 1,650

Show your work here

Go Online You can complete an Extra Example online.

Foldables It's time to update your Foldable, located in the Module Review, based on what you learned in this lesson. If you haven't already assembled your Foldable, you can find the instructions on page FL1.

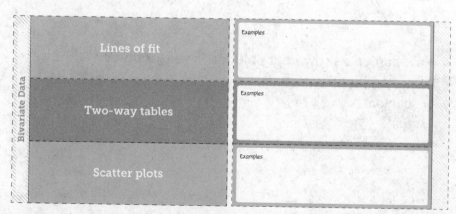

Practice

Go Online You can complete your homework online.

1. The scatter plot shows the number of girls that participated in high school sports. Write an equation in slope-intercept form for the line of fit that is drawn. Then interpret the slope and *y*-intercept. (Example 1)

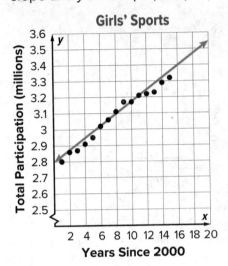

2. The scatter plot shows the tips different restaurant servers earned one night. Write an equation in slope-intercept form for the line of fit that is drawn. Then interpret the slope and *y*-intercept. (Example 1)

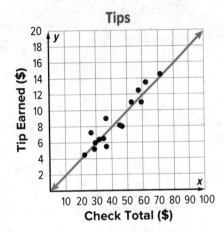

3. The scatter plot shows the relationship between the number of times a cricket chirps and the current temperature. Write an equation for the line of fit. Then use it to make a conjecture about the temperature when there are 40 cricket chirps. (Example 2)

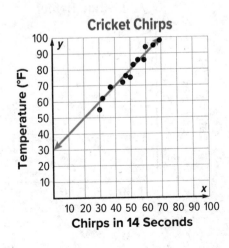

Test Practice

4. Multiple Choice The scatter plot shows the results of a survey about age and daily time spent playing video games. Which equation best represents the line of fit?

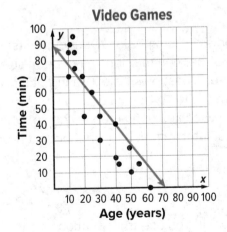

Ⓐ $y = 0.8x + 90$ Ⓒ $y = 1.25x + 90$

Ⓑ $y = -0.8x + 90$ Ⓓ $y = -1.25x + 90$

Apply

5. The scatter plot shows the average ticket price and the number of wins for certain NFL teams. How much more is the average price of a ticket for a team with 14 wins than a team with 3 wins? Round to the nearest dollar if necessary.

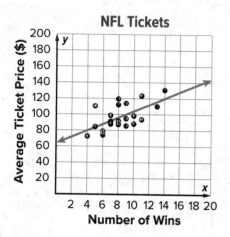

6. The scatter plot shows the number of T-shirts sold at different prices in a souvenir shop. How many more T-shirts were sold for $9 than $16?

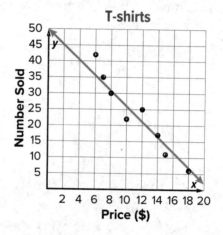

7. When will the slope of a line of fit be positive? When will it be negative?

8. **MP** **Make a Conjecture** What would make a scatter plot and its corresponding line of fit more useful to make accurate predictions? Will a line of fit always predict what will happen in the future? Explain your reasoning.

9. Lisle works on commission. During the first six months of the year her commissions have been steadily falling. The scatter plot shows Lisle's eating habits during these six months, with "1" being January.

   a. Do the two lines of fit intersect? If so, what is the point of intersection and what does it represent?

   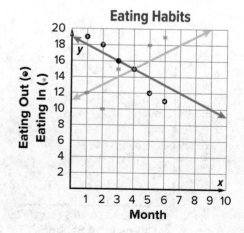

   b. How could you use the equations for the lines of fit to verify that the point (4, 15) is common to both lines?

Two-Way Tables

I Can... construct and interpret a two-way table using relative frequencies.

Learn Construct Two-Way Tables

A **two-way table** shows data from one sample as it relates to two different categories. One category is represented by rows, and the other is represented by columns.

Go Online Watch the animation to learn more about two-way tables.

The animation shows the following two-way table containing the results of a survey about drama and art participation. One hundred forty-six students were surveyed.

	Art	No Art	Total
Drama	26	38	64
No Drama	68	14	82
Total	94	52	146

Each row shows the frequency of student participation in drama. Each column shows the frequency of student participation in art. Twenty-six students participate in both art and drama. Fourteen students do not participate in either art or drama.

How many students participate in drama, but not art? _____

How many students participate in art, but not drama? _____

The last number in each row is the sum or total of the frequencies in that row. A total of 64 students participate in drama, while a total of 82 students do not participate in drama.

The last number in each column is the sum or total of the frequencies in that column. A total of 94 students participate in art, while a total of 52 students do not participate in art.

(continued on next page)

What Vocabulary Will You Learn?
relative frequency

two-way table

Talk About It!
Is there another way you can represent the information, other than by using a two-way table? Explain.

The animation also shows how to complete a two-way table by using the sums of the rows and columns. The partially-completed two-way table shows the results of a survey in which students were asked if they own a skateboard and if they own a bike.

	Own a Skateboard	No Skateboard	Total
Own a Bike	42		89
No Bike		21	
Total	57		

A total of 89 students own a bike. If 42 also own a skateboard, how many of the students who own a bike do not own a skateboard?

$89 - 42 = 47$ Subtract. 47 students who own a bike do not own a skateboard.

A total of 57 students own a skateboard. If 42 also own a bike, how many of the students who own a skateboard do not own a bike?

$57 - 42 = 15$ Subtract. 15 students who own a skateboard do not own a bike.

The table is completed with the information that you now know.

	Own a Skateboard	No Skateboard	Total
Own a Bike	42	$89 - 42 = 47$	89
No Bike	$57 - 42 = 15$	21	
Total	57		125

Find the total number of students who do not own a skateboard

$47 + 21 = 68$ Add. 68 students do not own a skateboard.

Find the total number of students who do not own a bike.

$15 + 21 = 36$ Add. 36 students do not own a skateboard.

The table is completed with all of the missing information.

	Own a Skateboard	No Skateboard	Total
Own a Bike	42	47	89
No Bike	15	21	$15 + 21 = 36$
Total	57	$47 + 21 = 68$	125

Talk About It!

How can you verify that your completed two-way table is correct?

🌐 Example 1 Construct Two-Way Tables

Felipe surveyed students at his school. He found that 78 students own a cell phone and 57 of those students own a tablet. There are 70 students that own a tablet. Nine students do not own either device.

Construct a two-way table summarizing the data.

Create a table using the two categories and the given values.

	Tablet	No Tablet	Total
Cell Phone	57		78
No Cell Phone		9	
Total	70		

A total of 78 students own a cell phone. How many students who own a cell phone do not own a tablet?

$78 - 57 =$ _____   Subtract. Complete the table above with this information.

A total of 70 students own a tablet. How many students who own a tablet do not own a cell phone?

$70 - 57 =$ _____   Subtract. Complete the table above with this information.

Find the total number of students who do not own a tablet.

$21 + 9 =$ _____   Add. Complete the table above with this information.

Find the total number of students who do not own a cell phone.

$13 + 9 =$ _____ .   Add. Complete the table above with this information.

Find the total number of students surveyed. You can either add 78 and 22 to obtain 100, or you can add 70 and 30 to obtain 100. So, 100 students were surveyed. Complete the table above with this information.

Check

Eloise surveyed the students in her cafeteria and found that 38 males agree with the new cafeteria rules while 70 do not. There were 92 females surveyed and 41 of them agree with the new cafeteria rules. Complete the two-way table to represent the data.

	Agree	Disagree	Total
Males			
Females			
Total			

🐾 **Go Online** You can complete an Extra Example online.

🗪 Talk About It!

What conclusions, if any, can you make from the data in the two-way table?

Learn Find and Interpret Relative Frequencies in Two-Way Tables

A two-way table can show ratios, or relative frequencies, for rows or for columns, rather than the actual values. **Relative frequency** is the ratio of the value of a subtotal to the value of the total. Relative frequencies can be written as ratios, decimals, or percents. In this lesson, relative frequencies are written as decimals.

	Play a Sport	Do Not Play a Sport	Total
On the Honor Roll	250	115	365
Not On the Honor Roll	45	30	75
Total	295	145	440

The following tables demonstrate the different types of relative frequencies that can be displayed in a two-way table.

To find the relative frequencies by column, divide each value by the total for that column. Round to the nearest hundredth if necessary.

	Play a Sport	Do Not Play a Sport	Total
On the Honor Roll	$250; \frac{250}{295} \approx 0.85$	$115; \frac{115}{145} \approx 0.79$	
Not On the Honor Roll	$45; \frac{45}{295} \approx 0.15$	$30; \frac{30}{145} \approx 0.21$	
Total	$295; \frac{295}{295} = 1.00$	$145; \frac{145}{145} = 1.00$	

To find the relative frequencies by row, divide each value by the total for that row. Round to the nearest hundredth if necessary.

	Play a Sport	Do Not Play a Sport	Total
On the Honor Roll	$250; \frac{250}{365} \approx 0.68$	$115; \frac{115}{365} \approx 0.32$	$365; \frac{365}{365} = 1.00$
Not On the Honor Roll	$45; \frac{45}{75} \approx 0.60$	$30; \frac{30}{75} \approx 0.40$	$75; \frac{75}{75} = 1.00$
Total			

You can use the relative frequencies to interpret the data. In the table above, 0.68, or 68%, of the students on the Honor Roll also play a sport. The table also shows that 0.60, or 60%, of the students not on the Honor Roll also play a sport, 0.32, or 32%, of the students on the Honor Roll do not play a sport, and 0.40, or 40%, of the students not on the Honor Roll do not play a sport.

💬 Talk About It!

Refer to the relative frequencies and totals for the columns. How can you interpret the data?

🌐 Example 2 Find and Interpret Row Relative Frequencies

The table shows the results of a survey that asked adults and students at a school whether they preferred to use a tablet or a laptop.

	Tablet	Laptop	Total
Students	57	21	78
Adults	13	9	22
Total	70	30	100

Find the relative frequencies by row. Are students or adults more likely to prefer laptops? Explain.

Part A Find the relative frequencies by row.

You need to compare students and adults in relation to laptops. Because students and adults are grouped by rows, find the row relative frequencies. Write the ratios of each value to the total in that row. Then write the ratio as a decimal. Round to the nearest hundredth.

	Tablet	Laptop	Total
Students	$57; \frac{57}{78} \approx$ ☐	$21; \frac{21}{78} \approx$ ☐	78; 1.00
Adults	$13; \frac{13}{22} \approx$ ☐	$9; \frac{9}{22} \approx$ ☐	22; 1.00

Part B Determine if students or adults are more likely to prefer laptops. Explain.

The row relative frequencies are shown.

	Tablet	Laptop	Total
Students	0.73	0.27	1.00
Adults	0.59	0.41	1.00

Study the relative frequencies for laptops. Adults are more likely than students to prefer laptops, because 0.41 is greater than 0.27.

💭 Think About It!

What value will you use for the total for the first row? the second row?

💬 Talk About It!

In order to determine if students or adults prefer laptops more, why should you examine the relative frequencies instead of the number of responses?

Check

The two-way table shows data from a survey about the number of students that agree or disagree with the new cafeteria rules. Find the relative frequencies by row. Are males or females more likely to disagree with the new rules? Explain.

Part A

Find the relative frequencies by row. Round to the nearest hundredth if necessary.

	Agree	Disagree	Total
Males	38;	70;	108; 1.00
Females	41;	51;	92; 1.00

Part B

Are males or females more likely to disagree with the new rules? Explain.

 Go Online You can complete an Extra Example online.

Pause and Reflect

Did you make any errors when completing the Check exercise? What can you do to make sure you don't repeat that error in the future?

> Record your observations here

🌐 Example 3 Find and Interpret Column Relative Frequencies

The table shows the results of a survey about the number of students that are left-handed or right-handed and whether they prefer music or sports.

	Left-Handed	Right-Handed	Total
Music	8	55	63
Sports	7	80	87
Total	15	135	150

Find the relative frequencies by column. Is a left-handed student or a right-handed student more likely to prefer music? Explain.

Part A Find the relative frequencies by column.

You need to compare left-handed students to right-handed students in relation to music. Because left-handed students and right-handed students are grouped by column, find the column relative frequencies. Write the ratios of each value to the total in that column. Then write the ratio as a decimal. Round to the nearest hundredth.

	Left-Handed	Right-Handed
Music	$8; \frac{8}{15} \approx \boxed{}$	$55; \frac{55}{135} \approx \boxed{}$
Sports	$7; \frac{7}{15} \approx \boxed{}$	$80; \frac{80}{135} \approx \boxed{}$
Total	15; 1.00	135; 1.00

Part B Determine if a left-handed student or a right-handed student is more likely to prefer music. Explain.

The column relative frequencies are shown.

	Left-Handed	Right-Handed
Music	0.53	0.41
Sports	0.47	0.59
Total	1.00	1.00

Study the relative frequencies for music. A left-handed student is more likely to prefer music than a right-handed student, because 0.53 is greater than 0.41.

😮 Think About It!

What value will you use for the total for the first column? the second column?

💬 Talk About It!

The table shows the column relative frequencies for the results of a survey about whether students or adults prefer chocolate or vanilla ice cream. How could you interpret the results of the survey based on the column relative frequencies?

	Students	Adults
Choc.	0.62	0.85
Van.	0.38	0.15
Total	1.00	1.00

Check

The two-way table shows data from a survey about the number of students enrolled in band and art class. Find the relative frequencies by column. Is a student that is enrolled in band or not enrolled in band more likely to be enrolled in art class? Explain.

Part A

Find the relative frequencies by column. Round to the nearest hundredth if necessary.

	Band	No Band
Art Class	15;	29;
No Art Class	33;	23;
Total	48; 1.00	52; 1.00

Part B

Is a student that is enrolled in band or not enrolled in band more likely to be enrolled in art class? Explain.

⬤ **Go Online** You can complete an Extra Example online.

📖 **Foldables** It's time to update your Foldable, located in the Module Review, based on what you learned in this lesson. If you haven't already assembled your Foldable, you can find the instructions on page FL1.

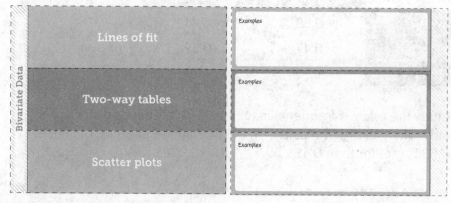

Practice

Go Online You can complete your homework online.

1. Omar surveyed students at his school. He found that 23 students are in the Chess Club, and 8 of those students are in the Math Club. There are 19 students that are in the Math Club. Ten students are in neither club. Construct a two-way table summarizing the data. (Example 1)

	Math Club	No Math Club	Total
Chess Club			
No Chess Club			
Total			

2. The table shows the results of a survey that asked seventh and eighth grade students whether they buy or pack their lunch. Find the relative frequencies. Round to the nearest hundredth. Are seventh graders or eighth graders more likely to buy their lunch? Explain. (Example 2)

	Buy Lunch	Pack a Lunch	Total
7th Graders	30	45	75
8th Graders	51	25	76
Total	81	70	151

3. The table shows the results of a survey about the number of bus riders at McGuffey Junior High. Find the relative frequencies. Round to the nearest hundredth. Are male students or female students more likely to not ride the bus? Explain. (Example 3)

	Male	Female	Total
Bus	110	84	194
No Bus	85	42	127
Total	195	126	321

4. **Multiselect** The two-way table shows the enrollment in language classes at Carson Middle School. Which of the following are valid conclusions about the data? Select all that apply.

	Enrolled in Spanish	Not Enrolled in Spanish	Total
Enrolled in French	30	65	95
Not Enrolled in French	20	5	25
Total	50	70	120

☐ Of the students that are enrolled in French, fewer than half of them are also enrolled in Spanish.

☐ More than half of the students are not enrolled in French or Spanish.

☐ Students that are enrolled in Spanish are likely to be enrolled in French.

☐ More than half of the students are enrolled in French.

☐ Students are more likely to be enrolled in Spanish than not in Spanish.

Apply

5. The Venn diagram shows data about the roller coasters at two different amusement parks. It compares whether or not they are sit-down coasters and whether or not they have loops. Construct a two-way table to represent the data.

	Sit-Down	Suspended	Total
Loops			
No Loops			
Total			

Sit-Down Versus Suspended Roller Coasters

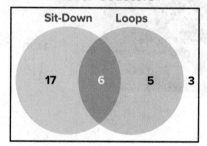

6. 🅜🅟 **Find the Error** Natalia surveyed 150 people about when going to the movie theater, if they like to watch comedies or dramas and whether or not they buy popcorn. Out of 90 people that liked comedies, 75 said they buy popcorn. There were 40 people that said they do not buy popcorn. Natalia created the two-way table shown to display the results. Find her mistake and correct it.

	Popcorn	No Popcorn	Total
Comedy	75	35	110
Drama	15	25	40
Total	90	60	150

Associations in Two-Way Tables

I Can... use relative frequencies to determine if an association exists between categories in a two-way table.

Explore Patterns of Association in Two-Way Tables

Online Activity You will explore how to find associations in two-way tables.

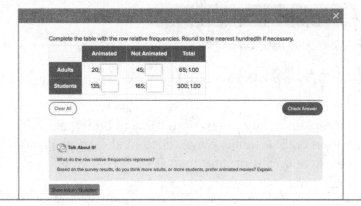

Complete the table with the row relative frequencies. Round to the nearest hundredth if necessary.

	Animated	Not Animated	Total
Adults	20;	45;	65; 1.00
Students	135;	165;	300; 1.00

Clear All   Check Answer

Talk About It!

What do the row relative frequencies represent?

Based on the survey results, do you think more adults, or more students, prefer animated movies? Explain.

Show Inquiry Question

Pause and Reflect

Did you struggle with any of the concepts in this Explore? How do you feel when you struggle with math concepts? What steps can you take to understand those concepts?

> Record your observations here

The row relative frequencies for the two-way table are shown. The greater the difference between the relative frequencies, the stronger the association. Explain why this is the case.

	Soccer	Tennis	Total
M	20; 0.25	60; 0.75	80; 1.00
F	95; 0.79	25; 0.21	120; 1.00

Learn Associations in Two-Way Tables

You can determine if data suggest that an association exists between two categories in a two-way table by examining the row or column relative frequencies.

The two-way table shows a difference in row relative frequencies between males (M) and females (F) and the sport they prefer. Since the relative frequencies for males are different than the relative frequencies for females, the data suggest that there is an association between gender and sport preference.

Row Relative Frequencies

	Soccer	Tennis	Total
Male	20; 0.25	60; 0.75	80; 1.00
Female	66; 0.55	54; 0.45	120; 1.00

A soccer player chosen at random is more likely to be female since 0.55 is greater than 0.25. A tennis player chosen at random is more likely to be male since 0.75 is greater than 0.45.

The column relative frequencies for a random survey about whether or not males and females play an instrument and whether or not males and females play a sport are in the two-way table shown. Since the relative frequencies for those that play an instrument are the same as those that play a sport, the data suggest that there is no association between what a person plays and their gender.

Column Relative Frequencies

	Play an Instrument	Play a Sport
Male	28; 0.40	52; 0.40
Female	42; 0.60	78; 0.60
Total	70; 1.00	130; 1.00

A male student chosen at random is just as likely to play an instrument as he is to play a sport. A female student chosen at random is just as likely to play an instrument as she is to play a sport.

Example 1 Use Row Relative Frequencies to Determine Associations

The two-way table shows how people of different ages get their news.

Find the row relative frequencies. Then determine if the data suggest an association between the categories. Explain your reasoning.

	TV	Internet	Total
Under 18	20	70	90
18 and Older	85	30	115
Total	105	100	205

Part A Find the row relative frequencies.

Calculate the row relative frequencies. Round to the nearest hundredth if necessary.

	TV	Internet	Total
Under 18	20; ☐	70; ☐	90; 1.00
18 and Older	85; ☐	30; ☐	115; 1.00

Part B Determine if the data suggest an association between the categories and explain your reasoning.

	TV	Internet	Total
Under 18	0.22	0.78	1.00
18 and Older	0.74	0.26	1.00

The data suggest that there is an association between age and how people get their news, because the relative frequencies are different. A person chosen at random that gets their news from the Internet is more likely to be younger than 18 than 18 and older.

Think About It!

How can you determine if the data suggest an association between the categories?

Talk About It!

Suppose you were to find the column relative frequencies instead of the row relative frequencies. Can you still determine that the data suggest an association between the categories? Explain.

Check

The two-way table shows the number of students that prefer a particular movie genre. Find the row relative frequencies. Then determine if that data suggest an association between the categories. Explain your reasoning.

	Comedy	Drama	Total
7th grade	27	57	84
8th grade	31	65	96
Total	58	122	180

Part A

Find the row relative frequencies. Round to the nearest hundredth if necessary.

	Comedy	Drama	Total
7th grade	27; ☐	57; ☐	84; 1.00
8th grade	31; ☐	65; ☐	96; 1.00

Part B

Determine if the data suggest an association between the categories. Explain your reasoning.

🔊 **Go Online** You can complete an Extra Example online.

🌐 Example 2 Use Column Relative Frequencies to Determine Associations

The two-way table shows the grade level for a student and the subject they prefer.

Find the column relative frequencies. Then determine if the data suggest an association between the categories. Explain your reasoning.

	Math	Language Arts	Total
7th grade	173	79	252
8th grade	102	46	148
Total	275	125	400

😮 **Think About It!**

How can you determine if the data suggest an association between the categories?

Part A Find the column relative frequencies.

Calculate the column relative frequencies. Round to the nearest hundredth if necessary.

	Math	Language Arts
7th grade	173; ☐	79; ☐
8th grade	102; ☐	46; ☐
Total	275; 1.00	125; 1.00

💬 **Talk About It!**

If you were to calculate the row relative frequencies instead of the column relative frequencies, does the association change? Explain.

Part B Determine if the data suggest an association between the categories. Explain your reasoning.

	Math	Language Arts
7th grade	0.63	0.63
8th grade	0.37	0.37
Total	1.00	1.00

The data suggest that there is no association between subject preference and grade level, because the relative frequencies are the same. A person chosen at random that is in 7th grade is just as likely to prefer Math as they are to prefer Language Arts.

Check

The two-way table shows whether or not someone owns a pet and the number of children in their family. Find the column relative frequencies. Then determine if the data suggest an association between the categories. Explain your reasoning.

	2 Children or Less	More Than 2 Children	Total
Owns a Pet	55	15	70
Does Not Own a Pet	10	20	30
Total	65	35	100

Part A

Find the column relative frequencies. Round to the nearest hundredth if necessary.

	2 Children or Less	More Than 2 Children
Owns a Pet	55; ☐	15; ☐
Does Not Own a Pet	10; ☐	20; ☐
Total	65; 1.00	35; 1.00

Part B

Determine if the data suggest an association between the categories. Explain your reasoning.

Go Online You can complete an Extra Example online.

🌐 Apply Jobs

Martin surveyed 150 tenth-grade students to find out if they have a part-time job. There are 94 students who have a part-time job, including 37 that also play a sport. Half of the students who do not have a job, play a sport. Do the data suggest that there is an association between having a job and playing a sport? Explain.

🔾 **Go Online**
Watch the animation.

Students without a job	Students with a job
56	▶ 94

1 What is the task?

Make sure you understand exactly what question to answer or problem to solve. You may want to read the problem three times. Discuss these questions with a partner.

First Time Describe the context of the problem, in your own words.
Second Time What mathematics do you see in the problem?
Third Time What are you wondering about?

2 How can you approach the task? What strategies can you use?

Record your observations here

3 What is your solution?

Use your strategy to solve the problem.

Show your work here

🗨 **Talk About It!**
Explain the process you used to solve the problem.

4 How can you show your solution is reasonable?

✏ **Write About It!** Write an argument that can be used to defend your solution.

Check

A group of two hundred 21-year-olds were surveyed about whether they live with their parents and if they attend college. There are eighty 21-year-olds that live with their parents. There are one hundred sixty 21-year-olds that attend college, and ninety-six do not live with their parents. Is there an association between attending college and living with parents for 21-year-olds? Explain.

Show your work here

🔲 **Go Online** You can complete an Extra Example online.

📖 **Foldables** It's time to update your Foldable, located in the Module Review, based on what you learned in this lesson. If you haven't already assembled your Foldable, you can find the instructions on page FL1.

Bivariate Data		Examples
	Lines of fit	
	Two-way tables	Examples
	Scatter plots	Examples

Practice

🔄 **Go Online** You can complete your homework online.

1. The two-way table shows the number of seventh and eighth grade students that plan on attending the school dance. Find the row relative frequencies. Then determine if the data suggest an association between the categories. Explain your reasoning.
(Example 1)

	Seventh	Eighth	Total
Attending	80; ☐	138; ☐	218; ☐
Not Attending	105; ☐	97; ☐	202; ☐
Total	185	235	420

2. The two-way table shows the results of a survey about two possible new art classes to be offered at the community center. Find the column relative frequencies. Then determine if the data suggest an association between the categories. Explain your reasoning. (Example 2)

	Pottery	Photography	Total
Under 30	43; ☐	86; ☐	129
30 and Older	66; ☐	55; ☐	121
Total	109; ☐	141; ☐	250

Test Practice

3. **Multiple Choice** The two-way table shows the number of middle school and high school students that use social media.

Based on the relative frequencies, which one of the following is *not* true?

	Social Media	No Social Media	Total
Middle School	410	815	1,225
High School	1,310	440	1,750
Total	1,720	1,255	2,975

Ⓐ A student that is chosen at random that uses social media is more likely to be a middle school student than a high school student.

Ⓑ A student that is chosen at random that does not use social media is more likely to be a middle school student than a high school student.

Ⓒ A middle school student that is chosen at random is less likely to use social media.

Ⓓ A high school student that is chosen at random is more likely to use social media.

Apply

4. Ninety people in a store were asked whether they liked jeans or khakis and whether they liked T-shirts or tank tops. Out of 25 people that liked khakis, 15 liked T-shirts. There were 55 customers that liked T-shirts. Is there an association between liking jeans and T-shirts? Explain.

5. The 275 eighth grade students at Hamilton Middle School went to an amusement park for their end-of-year celebration. Out of the 205 students that rode a roller coaster, 48 did not go on a water ride. There were 14 students that did not go on a roller coaster or a water ride. Is there an association between riding a roller coaster but not a water ride? Explain.

6. Use your own words to explain how you determine whether or not there is an association between the data in a two-way table.

7. **Create** Write a real-world problem in which you could represent the data with a two-way table. Then determine whether or a not the data suggest an association.

8. **MP Identify Structure** When constructing a two-way table, does it matter which categories are listed across the top row and which ones are listed on the side? Explain.

9. **MP Make A Conjecture** The results of a survey about what electronic technology customers in a store use are displayed in the two-way table.

	E-Reader	No E-Reader	Total
Laptop	20	8	28
No Laptop	45	22	67
Total	65	30	95

How could you use the information in the table to make a prediction about the number of people in a group of 200 that use both a laptop and an E-reader?

📖 **Foldables** Use your Foldable to help review the module.

Bivariate Data

A Line of Fit is useful for:

A Two-Way Table is useful for:

A Scatter Plot is useful for:

Rate Yourself! ⬛ ◆ ★

Complete the chart at the beginning of the module by placing a checkmark in each row that corresponds with how much you know about each topic after completing this module.

Write about one thing you learned.

_____

_____

_____

Write about a question you still have.

_____

_____

_____

Reflect on the Module

Use what you learned about scatter plots and two-way tables to complete the graphic organizer.

e Essential Question

What do patterns in data mean and how are they used?

Scatter Plots	Two-Way Tables
What types of variable associations can be found in scatter plots?	What is relative frequency?
How are patterns used when analyzing data?	How are patterns used when analyzing data?

Test Practice

1. Grid Consider the data provided in the table regarding hot chocolate sales.
(Lesson 1)

Temperature (°Fahrenheit)	Servings Sold
32	225
54	117
28	222
8	315
22	254
42	169
12	300

A. Construct a scatter plot that displays the relationship between the number of servings of hot chocolate sold, and the outdoor temperature at a football stadium.

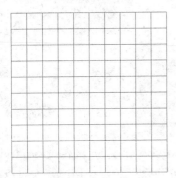

B. Which is the best interpretation of your scatter plot?

Ⓐ negative association; nonlinear

Ⓑ negative association; linear

Ⓒ positive association; nonlinear

Ⓓ positive association; linear

2. Open Response The scatter plot shows the amount of money earned (in tips) by different hairdressers, and the number of clients they served in a week. (Lesson 2)

A. Describe the association between the variables, and assess the line of fit.

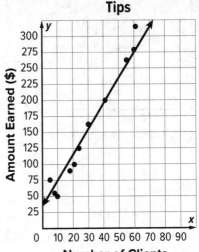

B. Use the line of fit to complete the conjecture.
We can predict that a hairdresser will earn about $____ if they serve 10 clients in a week.

3. Open Response The scatter plot shows the outdoor temperature and the number of guests that ride the water ride. Use the line of fit to make a conjecture about the number of riders if the temperature is 70°F. (Lesson 2)

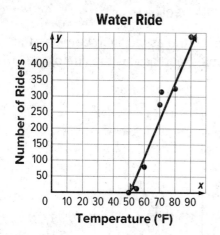

Water Ride

[Scatter plot with x-axis "Temperature (°F)" ranging 0 to 90 and y-axis "Number of Riders" ranging 0 to 450]

4. Multiselect The scatter plot shows the amount of time different students spent practicing for a typing speed test and their score, in words per minute. Which of the following statements accurately describe the situation? Select all that apply. (Lesson 3)

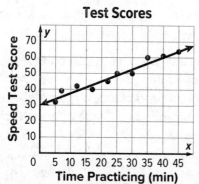

Test Scores

[Scatter plot with x-axis "Time Practicing (min)" ranging 0 to 45 and y-axis "Speed Test Score" ranging 0 to 70]

☐ The equation $y = \frac{3}{4}x + 30$ represents the line of fit.

☐ The equation $y = \frac{-3}{4}x + 30$ represents the line of fit.

☐ A student will receive a score of about 68 if they do not practice.

☐ A student will receive a score of about 30 if they do not practice.

☐ The y-intercept of the line is 0.

5. Table Item Mr. Edwards surveyed a total of 80 students in his first three classes in order to determine the numbers of students that have pets. Out of the 30 students in his 1st period class, 12 have no pets. Eighteen of the 27 students in his 2nd period class do have pets. Fourteen of the students in his 3rd period class do have pets. Construct a two-way table summarizing the data. (Lesson 4)

6. Multiple Choice The two-way table shows how people of different ages prefer to communicate. Which one of the statements accurately describes the association, if any, between age and how people prefer to communicate? (Lesson 5)

	Phone Call	Text Message	Total
Under 50	5	90	95
50 and Older	60	30	90
Total	65	120	185

Ⓐ A person chosen at random that communicates by phone call is more likely to be younger than 50, than 50 and older.

Ⓑ A person chosen at random that communicates by text message is more likely to be 50 and older, than younger than 50.

Ⓒ A person chosen at random that communicates by text message is more likely to be younger than 50, than 50 and older.

Ⓓ There does not appear to be an association between age and how people prefer to communicate.

📖 Foldables Study Organizers

What Are Foldables and How Do I Create Them?

Foldables are three-dimensional graphic organizers that help you create study guides for each module in your book.

Step 1 Go to the back of your book to find the Foldable for the module you are currently studying. Follow the cutting and assembly instructions at the top of the page.

Step 2 Go to the Module Review at the end of the module you are currently studying. Match up the tabs and attach your Foldable to this page. Dotted tabs show where to place your Foldable. Striped tabs indicate where to tape the Foldable.

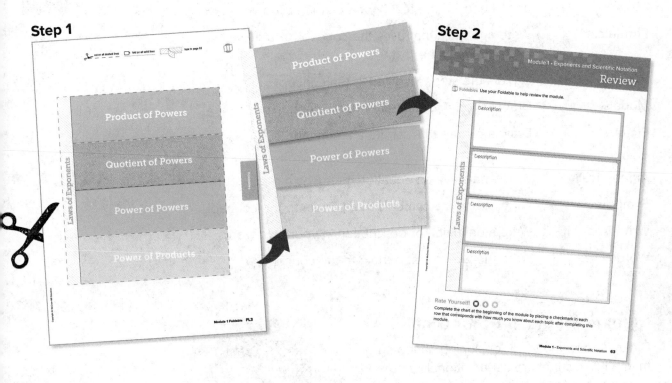

How Will I Know When to Use My Foldable?

You will be directed to work on your Foldable at the end of selected lessons. This lets you know that it is time to update it with concepts from that lesson. Once you've completed your Foldable, use it to study for the module test.

How Do I Complete My Foldable?

No two Foldables in your book will look alike. However, some will ask you to fill in similar information. Below are some of the instructions you'll see as you complete your Foldable. **HAVE FUN** learning math using Foldables!

Instructions and What They Mean

Best Used to...	Complete the sentence explaining when the concept should be used.
Definition	Write a definition in your own words.
Description	Describe the concept using words.
Equation	Write an equation that uses the concept. You may use one already in the text or you can make up your own.
Example	Write an example about the concept. You may use one already in the text or you can make up your own.
Formulas	Write a formula that uses the concept. You may use one already in the text.
How do I ...?	Explain the steps involved in the concept.
Models	Draw a model to illustrate the concept.
Picture	Draw a picture to illustrate the concept.
Solve Algebraically	Write and solve an equation that uses the concept.
Symbols	Write or use the symbols that pertain to the concept.
Write About It	Write a definition or description in your own words.
Words	Write the words that pertain to the concept.

Meet Foldables Author Dinah Zike

Dinah Zike is known for designing hands-on manipulatives that are used nationally and internationally by teachers and parents. Dinah is an explosion of energy and ideas. Her excitement and joy for learning inspires everyone she touches.

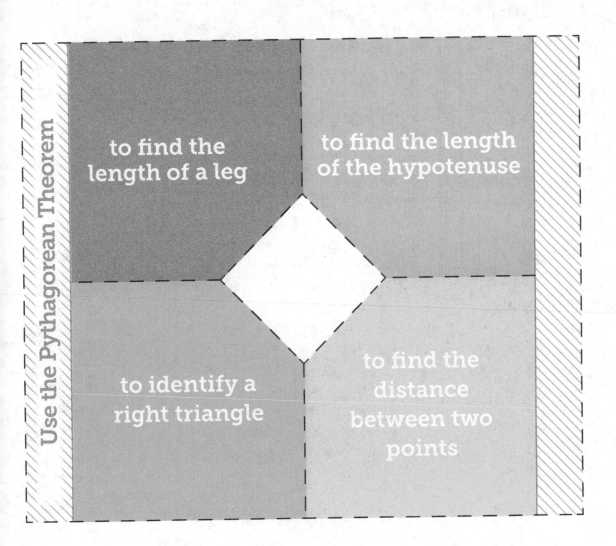

Use the Pythagorean Theorem

to find the
length of a leg

to find the length
of the hypotenuse

to identify a
right triangle

to find the
distance
between two
points

Foldables

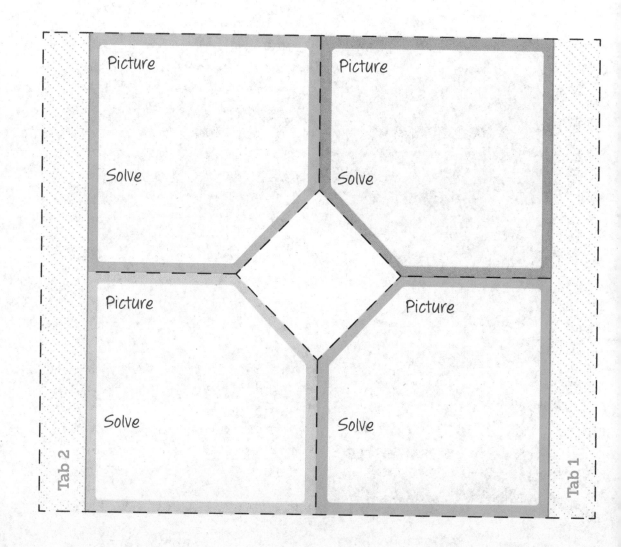

Transformations

translation	reflection
rotation	dilation

Symbols

Symbols

Symbols

Symbols

Tab 2

Tab 1

Congruent Figures

attributes	transformations
attributes	transformations

Similar Figures

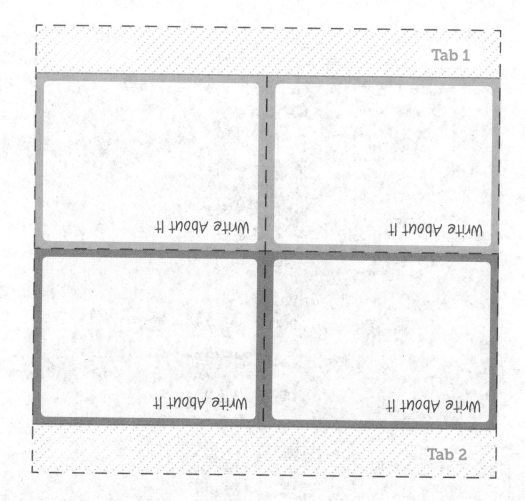

Tab 1

Write About It

Write About It

Write About It

Write About It

Tab 2

Volume

| cylinders | cones | spheres |

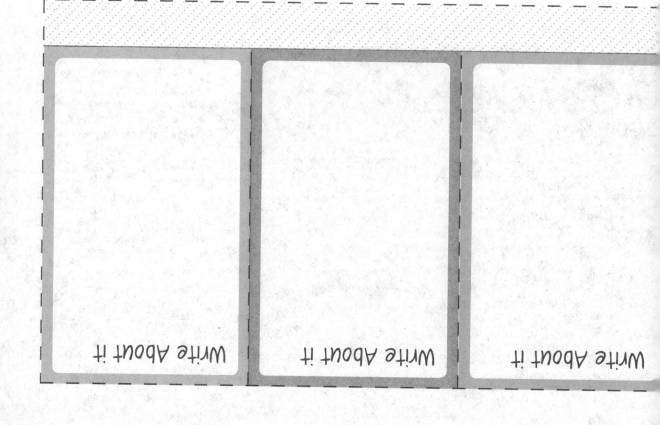

Write About it

Write About it

Write About it

cut on all dashed lines

fold on all solid lines

tape to page 629

Bivariate Data

Lines of fit

Two-way tables

Scatter plots

Foldables

Examples

Examples

Examples

Glossary

The Multilingual eGlossary contains words and definitions in the following 14 languages:

Arabic	English	Hmong	Russian	Urdu
Bengali	French	Korean	Spanish	Vietnamese
Brazilian Portuguese	Haitian Creole	Mandarin	Tagalog	

English / Español

A

algebra (Lesson 9-1) A branch of mathematics that involves expressions with variables.

álgebra Rama de las matemáticas que trabaja con expresiones con variables.

alternate exterior angles (Lesson 7-1) Exterior angles that lie on opposite sides of the transversal.

ángulos alternos externos Ángulos externos que se encuentran en lados opuestos de la transversal.

alternate interior angles (Lesson 7-1) Interior angles that lie on opposite sides of the transversal.

ángulos alternos internos Ángulos internos que se encuentran en lados opuestos de la transversal.

angle of rotation (Lesson 8-3) The degree measure of the angle through which a figure is rotated.

ángulo de rotación Medida en grados del ángulo sobre el cual se rota una figura.

Angle-Angle Similarity (Lesson 9-4) If two angles of one triangle are congruent to two angles of another triangle, then the triangles are similar.

similitud ángulo-ángulo Si dos ángulos de un triángulo son congruentes con dos ángulos de otro triángulo, entonces los triángulos son similares.

B

bar notation (Lesson 2-1) In repeating decimals, the line or bar placed over the digits that repeat.

notación de barra Línea o barra que se coloca sobre los dígitos que se repiten en decimales periódicos.

base (Lesson 1-1) In a power, the number that is the common factor. In 10^3, the base is 10. That is, $10^3 = 10 \times 10 \times 10$.

base En una potencia, el número que es el factor común. En 10^3, la base es 10. Es decir, $10^3 = 10 \times 10 \times 10$.

bivariate data (Lesson 11-1) Data with two variables, or pairs of numerical observations.

datos bivariantes Datos con dos variables, o pares de observaciones numéricas.

C

center of dilation (Lesson 8-4) The center point from which dilations are performed.

centro de la homotecia Punto fijo en torno al cual se realizan las homotecias.

center of rotation (Lesson 8-3) A fixed point around which shapes move in a circular motion to a new position.

centro de rotación Punto fijo alrededor del cual se giran las figuras en movimiento circular alrededor de un punto fijo.

cluster (Lesson 11-1)   A collection of points that are close together in a scatter plot.

racimo   Una colección de puntos que están muy juntos en un diagrama de dispersión.

coefficient (Lesson 3-1)   The numerical factor of a term that contains a variable.

coeficiente   Factor numérico de un término que contiene una variable.

composite solid (Lesson 10-5)   An object made up of more than one solid.

sólido complejo   Cuerpo compuesto de más de un sólido.

composition of transformations (Lesson 9-1) The resulting transformation when a transformation is applied to a figure and then another transformation is applied to its image.

composición de transformaciones   Transformación que resulta cuando se aplica una transformación a una figura y luego se le aplica otra transformación a su imagen.

cone (Lesson 10-2)   A three-dimensional figure with one circular base connected by a curved surface to a single point.

cono   Una figura tridimensional con una circular base conectada por una superficie curva para un solo punto.

congruent (Lesson 9-1)   Having the same measure; if one image can be obtained from another by a sequence of rotations, reflections, or translations.

congruente   Que tienen la misma medida; si una imagen puede obtenerse de otra por una secuencia de rotaciones, reflexiones o traslaciones.

constant (Lesson 3-1)   A term without a variable.

constante   Término sin variables.

constant of proportionality (Lesson 4-4)   The constant ratio in a proportional linear relationship.

constante de proporcionalidad   La razón constante en una relación lineal proporcional.

constant of variation (Lesson 4-4)   A constant ratio in a direct variation.

constante de variación   Razón constante en una relación de variación directa.

constant rate of change (Lesson 4-1)   The rate of change between any two points in a linear relationship is the same or *constant*.

tasa constante de cambio   La tasa de cambio entre dos puntos cualesquiera en una relación lineal permanece igual o *constante*.

converse (Lesson 7-4)   The converse of a theorem is formed when the *if* and *then* statements are reversed.

recíproco   El recíproco de un teorema se forma cuando las declaraciones de *si* y *luego* se invierten.

converse of the Pythagorean Theorem (Lesson 7-4) A theorem that can be used to test whether a triangle is a right triangle. If the sides of the triangle have lengths a, b, and c, such that $c^2 = a^2 + b^2$, then the triangle is a right triangle.

el recíproco del teorema de Pitágoras   Un teorema que puede usarse para probar si un triángulo es un triángulo rectángulo. Si los lados del triángulo tienen longitudes a, b, y c, tales que $c^2 = a^2 + b^2$, entonces el triángulo es un triángulo rectángulo.

coordinate plane (Lesson 4-1)   A coordinate system in which a horizontal number line and a vertical number line intersect at their zero points.

plano de coordenadas   Sistema de coordenadas en que una recta numérica horizontal y una recta numérica vertical se intersecan en sus puntos cero.

corresponding angles (Lesson 7-1)   Angles that are in the same position on two parallel lines in relation to a transversal.

ángulos correspondientes   Ángulos que están en la misma posición sobre dos rectas paralelas en relación con la transversal.

corresponding parts (Lesson 4-3) Parts of congruent or similar figures that are in the same relative position.

partes correspondientes Partes de figuras congruentes o similares que están en la misma posición relativa.

counterexample (Lesson 2-3) A statement or example that shows a conjecture is false.

contraejemplo Ejemplo o enunciado que demuestra que una conjetura es falsa.

cube root (Lesson 2-2) One of three equal factors of a number. If $a^3 = b$, then a is the cube root of b. The cube root of 64 is 4 since $4^3 = 64$.

raíz cúbica Uno de tres factores iguales de un número. Si $a^3 = b$, entonces a es la raíz cúbica de b. La raíz cúbica de 64 es 4, dado que $4^3 = 64$.

cylinder (Lesson 10-1) A three-dimensional figure with two parallel congruent circular bases connected by a curved surface.

cilindro Una figura tridimensional con dos paralelas congruentes circulares bases conectados por una superficie curva.

D
---

dilation (Lesson 8-4) A transformation that enlarges or reduces a figure by a scale factor.

homotecia Transformación que produce la ampliación o reducción de una imagen por un factor de escala.

direct variation (Lesson 4-4) A relationship between two variable quantities with a constant ratio. A proportional linear relationship.

variación directa Relación entre dos cantidades variables con una razón constante. Una relación lineal proporcional.

E
---

elimination (Lesson 6-4) An algebraic method that can be used to find the exact solution of a system of equations by eliminating one of the variables.

eliminación Un método algebraico que se puede usar para encontrar la solución exacta de un sistema de ecuaciones mediante la eliminación de una de las variables.

evaluate (Lesson 1-1) To find the value of an expression.

evaluar Calcular el valor de una expresión.

exponent (Lesson 1-1) In a power, the number of times the base is used as a factor. In 10^3, the exponent is 3.

exponente En una potencia, el número de veces que la base se usa como factor. En 10^3, el exponente es 3.

exterior angle (Lesson 7-2) An angle between one side of a polygon and the extension of an adjacent side.

ángulo exterior Un ángulo entre un lado de un polígono y la extensión de un lado adyacente.

exterior angles (Lesson 7-1) The four outer angles formed by two lines cut by a transversal.

ángulo externo Los cuatro ángulos exteriores que se forman cuando una transversal corta dos rectas.

F
---

function (Lesson 5-1) A relation in which each member of the input is paired with exactly one member of the output.

función Relación en la que cada elemento de la entrada le corresponde exactamente con un único elemento de la salida.

function table (Lesson 5-2) A table organizing the input, rule, and output of a function.

tabla de funciones Tabla que organiza la regla de entrada y de salida de una función.

hemisphere (Lesson 10-3) One of two congruent halves of a sphere.

hipotenuse (Lesson 7-3) The side opposite the right angle in a right triangle.

hemisferio Una de dos mitades congruentes de una esfera.

hipotenusa El lado opuesto al ángulo recto de un triángulo rectángulo.

image (Lesson 8-1) The resulting figure after a transformation.

imagen Figura que resulta después de una transformación.

indirect measurement (Lesson 9-5) A technique using properties of similar polygons to find distances or lengths that are difficult to measure directly.

medición indirecta Técnica que usa las propiedades de polígonos semejantes para calcular distancias o longitudes difíciles de medir directamente.

initial value (Lesson 4-5) The starting value in a real-world situation in which an equation can be written. The y-intercept of a linear function.

valor inicial El valor inicial en una situación real en la que se puede escribir una ecuación. La intersección y de una función lineal.

input (Lesson 5-1) The set of x-coordinates in a relation.

entrada El conjunto de x-coordenadas en una relación.

integers (Lesson 2-1) The set of whole numbers and their opposites.

enteros El conjunto de números enteros y sus opuestos.

interior angle (Lesson 7-2) An angle inside a polygon.

ángulo interno Ángulo dentro de un polígono.

interior angles (Lesson 7-1) The four inside angles formed by two lines cut by a transversal.

ángulo interno Los cuatro ángulos internos formados por dos rectas intersecadas por una transversal.

inverse operations (Lesson 2-2) Pairs of operations that undo each other. Addition and subtraction are inverse operations. Multiplication and division are inverse operations.

peraciones inversas Pares de operaciones que se anulan mutuamente. La adición y la sustracción son operaciones inversas. La multiplicación y la división son operaciones inversas.

irrational number (Lesson 2-3) A number that cannot be expressed as the ratio $\frac{a}{b}$, where a and b are integers and $b \neq 0$.

números irracionales Número que no se puede expresar como la proporción $\frac{a}{b}$, donde a y b son enteros y $b \neq 0$.

legs (Lesson 7-3) The two sides of a right triangle that form the right angle.

catetos Los dos lados de un triángulo rectángulo que forman el ángulo recto.

like terms (Lesson 3-3)   Terms that contain the same variable(s) to the same powers.

términos semejantes   Términos que contienen la misma variable o variables elevadas a la misma potencia.

linear (Lesson 4-1)   To fall in a straight line.

lineal   Que cae en una línea recta.

linear equation (Lesson 4-1)   An equation with a graph that is a straight line.

ecuación lineal   Ecuación cuya gráfica es una recta.

linear function (Lesson 5-2)   A function in which the graph of the solutions forms a line.

función lineal   Función en la cual la gráfica de las soluciones forma una recta.

linear relationship (Lesson 4-1)   A relationship that has a straight-line graph.

relación lineal   Relación cuya gráfica es una recta.

line of fit (Lesson 11-2)   A line that is very close to most of the data points in a scatter plot.

línea de ajuste   Línea que más se acerca a la mayoría de puntos de los datos en un diagrama de dispersión.

line of reflection (Lesson 8-2)   The line over which a figure is reflected in a transformation.

línea de reflexión   Línea a través de la cual se refleja una figura en una transformación.

line segment (Lesson 7-2)   Part of a line containing two endpoints and all of the points between them.

segmento de línea   Parte de una línea que contiene dos extremos y todos los puntos entre ellos.

M

---

monomial (Lesson 1-2)   A number, a variable, or a product of a number and one or more variables.

monomio   Un número, una variable o el producto de un número por una o más variables.

N

---

natural numbers (Lesson 2-1)   The set of numbers used for counting.

números naturales   El conjunto de números utilizado para el recuento.

negative exponent (Lesson 1-4)   The result of repeated division used to represent very small numbers.

exponente negative   El resultado de la división repetida se utiliza para representar números muy pequeños.

nonlinear function (Lesson 5-5)   A function whose rate of change is not constant. The graph of a nonlinear function is not a straight line.

función no lineal   Función cuya tasa de cambio no es constante. La gráfica de una función no lineal no es una recta.

Order of Operations (Lesson 1-1) When evaluating expressions with powers or more than one operation, use these rules.

1. Simplify the expression inside the grouping symbols.

2. Evaluate all powers.

3. Perform multiplication and division in order from left to right.

4. Perform addition and subtraction in order from left to right.

ordered pair (Lesson 4-1) A pair of numbers used to locate a point in the coordinate plane. The ordered pair is written in this form: (*x*-coordinate, *y*-coordinate).

origin (Lesson 4-1) The point of intersection of the *x*-axis and *y*-axis in a coordinate plane.

outlier (Lesson 11-1) A data point that is distinctly separate from the rest of the data.

output (Lesson 5-1) The set of *y*-coordinates in a relation.

orden de operaciones Al evaluar expresiones con poderes o más de una operación, utilice estas reglas.

1. Simplifique la expresión dentro de los símbolos de agrupación.

2. Evaluar todos los poderes.

3. Realizar multiplicación y división en orden de izquierda a derecha.

4. Realizar la suma y resta en orden de izquierda a derecha.

par ordenado Par de números que se utiliza para ubicar un punto en un plano de coordenadas. Se escribe de la siguiente forma: (coordenada *x*, coordenada *y*).

origen Punto en que el eje *x* y el eje *y* se intersecan en un plano de coordenadas.

valor atípico Un punto de datos que está claramente separado del resto de los datos.

salida El conjunto de *y*-coordenadas en una relación.

parallel lines (Lesson 7-1) Lines in the same plane that never intersect or cross. The symbol || means parallel.

perfect cube (Lesson 2-2) A number whose cube root is an integer. 27 is a perfect cube because its cube root is 3.

perfect square (Lesson 2-2) A number whose square root is a whole number. 25 is a perfect square because its square root is 5.

perpendicular lines (Lesson 7-1) Two lines that intersect to form right angles.

power (Lesson 1-1) A product of repeated factors using an exponent and a base. The power 7^3 is read *seven to the third power,* or *seven cubed.*

rectas paralelas Rectas que yacen en un mismo plano y que no se intersecan. El símbolo || significa paralela a.

cubo perfecto Número cuya raíz cúbica es un número entero. 27 es un cubo perfecto porque su raíz cúbica es 3.

cuadrados perfectos Número cuya raíz cuadrada es un número entero. 25 es un cuadrado perfecto porque su raíz cuadrada es 5.

rectas perpendiculares Dos rectas que se intersecan formando ángulos rectos.

potencia Producto de factores repetidos con un exponente y una base. La potencia 7^3 se lee *siete a la tercera potencia* o *siete al cubo.*

Power of a Power Property (Lesson 1-3) A property that states to find the power of a power, multiply the exponents.

potencia de una propiedad de potencia Una propiedad que declara encontrar el poder de un poder, multiplicar los exponentes.

Power of a Product Property (Lesson 1-3) A property that states to find the power of a product, find the power of each factor and multiply.

potencia de una propiedad de producto Una propiedad que declara encontrar el poder de un producto, encuentra el poder de cada factor y se multiplica.

preimage (Lesson 8-1) The original figure before a transformation.

preimagen Figura original antes de una transformación.

principal square root (Lesson 2-2) The positive square root of a number.

raíz cuadrada principal La raíz cuadrada positiva de un número.

Product of Powers Property (Lesson 1-2) A property that states to multiply powers with the same base, add their exponents.

producto de la propiedad de los poderes Una propiedad que declara multiplicar poderes con la misma base, añade sus exponentes.

proof (Lesson 7-4) A logical argument in which each statement that is made is supported by a statement that is accepted as true.

prueba Argumento lógico en el cual cada enunciado hecho se respalda con un enunciado que se acepta como verdadero.

Pythagorean Theorem (Lesson 7-3) In a right triangle, the square of the length of the hypotenuse c is equal to the sum of the squares of the lengths of the legs a and b. $a^2 + b^2 = c^2$

Teorema de Pitágoras En un triángulo rectángulo, el cuadrado de la longitud de la hipotenusa es igual a la suma de los cuadrados de las longitudes de los catetos. $a^2 + b^2 = c^2$

Q

quadrants (Lesson 4-1) The four sections of the coordinate plane.

cuadrantes Las cuatro secciones del plano de coordenadas.

qualitative graph (Lesson 5-6) A graph used to represent situations that do not necessarily have numerical values.

gráfica cualitativa Gráfica que se usa para representar situaciones que no tienen valores numéricos necesariamente.

Quotient of Powers Property (Lesson 1-2) A property that states to divide powers with the same base, subtract their exponents.

propiedad del cociente de poderes Una propiedad que declara dividir poderes con la misma base, resta sus exponentes.

R

radical sign (Lesson 2-2) The symbol used to indicate a positive square root, $\sqrt{}$.

signo radical Símbolo que se usa para indicar una raíz cuadrada no positiva, $\sqrt{}$.

rate of change (Lesson 4-1) A rate that describes how one quantity changes in relation to another quantity.

tasa de cambio Una tasa que describe cómo cambia una cantidad en relación con otra cantidad.

rational numbers (Lesson 2-1)   Numbers that can be written as the ratio of two integers in which the denominator is not zero. All integers, fractions, mixed numbers, and percents are rational numbers.

números racionales   Números que pueden escribirse como la razón de dos enteros en los que el denominador no es cero. Todos los enteros, fracciones, números mixtos y porcentajes son números racionales.

real numbers (Lesson 2-3)   The set of rational numbers together with the set of irrational numbers.

números reales   El conjunto de números racionales junto con el conjunto de números irracionales.

reflection (Lesson 8-2)   A transformation where a figure is flipped over a line. Also called a flip.

reflexión   Transformación en la cual una figura se voltea sobre una recta. También se conoce como simetría de espejo.

relation (Lesson 5-1)   Any set of ordered pairs.

relación   Cualquier conjunto de pares ordenados.

relative frequency (Lesson 11-4)   The ratio of the number of experimental successes to the total number of experimental attempts.

frecuencia relativa   Razón del número de éxitos experimentales al número total de intentos experimentales.

remote interior angles (Lesson 7-2)   The angles of a triangle that are not adjacent to a given exterior angle.

ángulos internos no adyacentes   Ángulos de un triángulo que no son adya centes a un ángulo exterior dado.

repeating decimal (Lesson 2-1)   A decimal in which 1 or more digits repeat.

decimal periódico   Un decimal en el que se repiten 1 o más dígitos.

right triangle (Lesson 7-3)   A triangle with one right angle.

triángulo rectángulo   Triángulo con un ángulo recto.

rise (Lesson 4-2)   The vertical change between any two points on a line.

elevación   El cambio vertical entre cualquier par de puntos en una recta.

rotation (Lesson 8-3)   A transformation in which a figure is turned about a fixed point.

rotación   Transformación en la cual una figura se gira alrededor de un punto fijo.

run (Lesson 4-2)   The horizontal change between any two points on a line.

carrera   El cambio horizontal entre cualquier par de puntos en una recta.

S

scale factor (Lesson 8-4)   The ratio of the lengths of two corresponding sides of two similar polygons.

factor de escala   La razón de las longitudes de dos lados correspondientes de dos polígonos semejantes.

scatter plot (Lesson 11-1)   A graph that shows the relationship between bivariate data graphed as ordered pairs on a coordinate plane.

diagrama de dispersión   Gráfica que muestra la relación entre datos bivariados graficadas como pares ordenados en un plano de coordenadas.

scientific notation (Lesson 1-5)   A compact way of writing numbers with absolute values that are very large or very small. In scientific notation, 5,500 is 5.5×10^3.

notación científica   Manera abreviada de escribir números con valores absolutos que son muy grandes o muy pequeños. En notación científica, 5,500 es 5.5×10^3.

similar (Lesson 9-3) If one image can be obtained from another by a sequence of transformations and dilations.

similar Si una imagen puede obtenerse de otra mediante una secuencia de transformaciones y dilataciones.

similar figures (Lesson 4-3) Figures that have the same shape but not necessarily the same size.

figuras semejantes Figuras que tienen la misma forma, pero no necesariamente el mismo tamaño.

slope (Lesson 4-1) The rate of change between any two points on a line. The ratio of the rise, or vertical change, to the run, or horizontal change.

pendiente Razón de cambio entre cualquier par de puntos en una recta. La razón de la altura, o cambio vertical, a la carrera, o cambio horizontal.

slope-intercept form (Lesson 4-5) An equation written in the form $y = mx + b$, where m is the slope and b is the y-intercept.

forma pendiente intersección Ecuación de la forma $y = mx + b$, donde m es la pendiente y b es la intersección y.

slope triangles (Lesson 4-3) Right triangles that fall on the same line on the coordinate plane.

triángulos de pendiente Triángulos rectos que caen en la misma línea en el plano de coordenadas.

solution (of an equation) (Lesson 4-1) A value that makes an equation true.

solución (de una ecuación) Un valor que hace que una ecuación sea verdadera.

solution (of a system of equations) (Lesson 6-1) An ordered pair that is a solution of both equations.

solución (de un sistema de ecuaciones) Un par ordenado que es una solución de ambas ecuaciones.

solid (Lesson 10-1) A three-dimensional figure formed by intersecting planes.

sólido Figura tridimensional formada por planos que se intersecan.

sphere (Lesson 10-3) The set of all points in space that are a given distance from a given point called the center.

esfera Conjunto de todos los puntos en el espacio que están a una distancia dada de un punto dado llamado centro.

square root (Lesson 2-2) One of the two equal factors of a number. If $a^2 = b$, then a is the square root of b. A square root of 144 is 12 since $12^2 = 144$.

raíz cuadrada Uno de dos factores iguales de un número. Si $a^2 = b$, la a es la raíz cuadrada de b. Una raíz cuadrada de 144 es 12 porque $12^2 = 144$.

standard form (Lesson 1-5) Numbers written without exponents.

forma estándar Números escritos sin exponentes.

substitution (Lesson 6-3) An algebraic model that can be used to find the exact solution of a system of equations.

sustitución Modelo algebraico que se puede usar para calcular la solución exacta de un sistema de ecuaciones.

system of equations (Lesson 6-1) A set of two or more equations with the same variables.

sistema de ecuaciones Sistema de ecuaciones con las mismas variables.

term (Lesson 1-2)   Each part of an algebraic expression separated by an addition or subtraction sign.

terminating decimal (Lesson 2-1)   A decimal where the repeating digit is zero.

theorem (Lesson 7-3)   A statement or conjecture that can be proven.

transformation (Lesson 8-1)   An operation that maps a geometric figure, preimage, onto a new figure, image.

translation (Lesson 8-1)   A transformation that slides a figure from one position to another without turning.

transversal (Lesson 7-1)   A line that intersects two or more other lines.

triangle (Lesson 7-2)   A figure formed by three line segments that intersect only at their endpoints.

truncating (Lesson 2-4)   A process of approximating a decimal number by eliminating all decimal places past a certain point without rounding.

two-way table (Lesson 11-4)   A table that shows data from one sample that pertain to two different categories.

término   Cada parte de un expresión algebraica separada por un signo adición o un signo sustracción.

decimal finito   Un decimal donde el dígito que se repite es cero.

teorema   Un enunciado o conjetura que puede probarse.

transformación   Operación que convierte una figura geométrica, la pre-imagen, en una figura nueva, la imagen.

traslación   Transformación en la cual una figura se desliza de una posición a otra sin hacerla girar.

transversal   Recta que interseca dos o más rectas.

triángulo   Figura formada por tres segmentos de recta que se intersecan sólo en sus extremos.

truncando   Proceso de aproximación de un número decimal eliminando todos los decimales más allá de un cierto punto sin redondear.

tabla de doble entrada   Una tabla que muestra datos de una muestra que pertenecen a dos categorías diferentes.

unit rate (Lesson 4-1)   A rate in which the first quantity is compared to 1 unit of the second quantity.

unit ratio (Lesson 4-1)   A ratio in which the first quantity is compared to every 1 unit of the second quantity.

tasa unitaria   Una tasa en la que la primera cantidad se compara con 1 unidad de la segunda cantidad.

razón unitaria   Una relación en la que la primera cantidad se compara con cada 1 unidad de la segunda cantidad.

variable (Lesson 3-1)   A symbol, usually a letter, used to represent a number in mathematical expressions or sentences.

variable   Un símbolo, por lo general, una letra, que se usa para representar números en expresiones o enunciados matemáticos.

vertex (Lesson 7-2)   The point where the sides of an angle meet.

vértice   Punto donde se encuentran los lados.

vertical line test (Lesson 5-1)   A test used to see whether a graph is a function. If, for each value of x, a vertical line passes through no more than one point on the graph, then the graph represents a function. If the line passes through more than one point on the graph, it is not a function.

prueba de línea vertical   Una prueba utilizada para ver si un gráfico es una función. Si, para cada valor de x, una línea vertical no pasa más de un punto en el gráfico, entonces el gráfico representa una función. Si la línea pasa por más de un punto en el gráfico, no es una función.

volume (Lesson 10-1)   The measure of the space occupied by a solid. Standard units of measure are cubic units such as in^3 or ft^3.

volumen   Medida del espacio que ocupa un sólido. Unidades de medida estándar son unidades cúbicas tales como $pulg^3$ o $pies^3$.

W

whole numbers (Lesson 2-1)   The set of numbers used for counting and zero.

números enteros   El conjunto de números utilizado para contar y cero.

X

x-axis (Lesson 4-1)   The horizontal number line that helps to form the coordinate plane.

eje x La recta numérica horizontal que ayuda a formar el plano de coordenadas.

x-coordinate (Lesson 4-1)   The first number of an ordered pair.

coordenada x El primer número de un par ordenado.

x-intercept (Lesson 4-1)   The x-coordinate of the point where the line crosses the x-axis.

intersección x La coordenada x del punto donde cruza la gráfica el eje x.

Y

y-axis (Lesson 4-1)   The vertical number line that helps to form the coordinate plane.

eje y La recta numérica vertical que ayuda a formar el plano de coordenadas.

y-coordinate (Lesson 4-1)   The second number of an ordered pair.

coordenada y El segundo número de un par ordenado.

y-intercept (Lesson 4-5)   The y-coordinate of the point where the line crosses the y-axis.

intersección y La coordenada y del punto donde cruza la gráfica el eje y.

Z

Zero Exponent Rule (Lesson 1-4)   A rule that states that any nonzero number to the zero power is equivalent to 1.

regla de exponente cero   Una regla que establece que cualquier número diferente de cero a la potencia cero es equivalente a 1.

A

Algebra tiles, 129–130

Algebraic expressions
evaluating, 8
like terms, 145, 151–154

Alternate exterior angles, 383

Alternate interior angles, 383

Angles
alternate exterior, 383
alternate interior, 383
corresponding, 383
exterior, 382, 397–400
interior, 382, 393
missing measures of,
386–388, 395–396, 400
remote interior, 397–398
sum of in a triangle, 394

Angle-Angle (AA) Similarity,
515–516

Appropriate units, 50

B

Bar notation, 70

Base of a power, 3, 6
negative bases, 6

Bivariate data, 581

C

Center of dilation, 466

Center of rotation, 455

Cluster, 583

Composite solids, 567
volume of, 568–570

**Composition of
transformations,** 483

Cones, 544
missing dimensions of, 561
volume of, 543–546

Congruence statements, 494

Congruent, 483–490

Constant of proportionality,
213–214

Constant of variation, 213–214

Constant rate of change, 176

Converse, 417

**Converse of the Pythagorean
Theorem,** 417–420

Coordinate plane. *See also*
graphing
dilations on, 465–470
distance on, 423–424
reflections on, 445–452
rotations on, 455–460
translations on, 435–440

Corresponding angles, 383

Corresponding parts, 205,
493–495, 513–514

Counterexample, 96–98

Cube roots, 84–86
estimating, 104
solving equations, 86

Cylinders, 535
missing dimensions of,
559–560
volume of, 535–538

D

Decimals
as fractions, 73
as mixed numbers, 74

Dilations, 465–470

Direct variation, 213–220

Distance
with the Pythagorean
Theorem, 423–424

E

Elimination method, 351

Equations
multi–step, 145–148
number of solutions of, 159-164
using cube roots to solve, 86
using square roots to solve, 83

with rational coefficients,
132–134
with variables on each side,
129–134, 137–140
writing, 137–140, 151–154

Estimating
cube roots, 104
scientific notation, 49-50
square roots, 102–103
truncating, 105

Evaluate, 7

Exponent, 3
negative, 36
of zero, 33

Exterior angles, 382
of a triangle, 397–400

F

Factors, 3–4

Foldables®, 22, 30, 63, 76,
98, 123, 148, 169, 188, 222,
234, 247, 260, 270, 302, 313,
328, 348, 375, 414, 420, 426,
429, 442, 452, 462, 472, 475,
490, 498, 510, 520, 529, 540,
548, 556, 575, 588, 596, 606,
616, 626, 629

Fractions
as decimals, 71

Function table, 263–265

Functions, 254
as equations, 273–280,
298–300
as graphs, 257–260,
266–268, 294–296
as mapping diagrams,
254–255
as tables, 256–257, 263–265,
296–297
comparing properties of,
285–288
identifying linearity, 293–300
nonlinear, 293
vertical line test for, 257–258

Graphic organizers, 64, 124, 170, 248, 314, 376, 430, 476, 530, 576, 630

Graphing
    equations in slope-intercept form, 237–239
    horizontal lines, 239–240
    linear functions, 266–268
    proportional relationships, 179–182
    right triangles, 423–424
    scatter plots, 581–583, 593–594
    systems of equations, 320–326
    transformations, 436, 446–448, 456, 468–469, 485–486, 488, 502–503, 505–506
    vertical lines, 241–242

Hemisphere, 554
    volume of, 554

Hypotenuse of a right triangle, 405

Image, 435

Indirect measurement, 523–526

Initial value, 229, 273

Input, 253

Integers, 69

Interior angles, 382
    of a triangle, 393–395

Inverse operations, 83

Irrational numbers, 91–92, 101–106
    estimating, 101–106

Legs of a right triangle, 405

Like terms, 145, 151–154

Line of reflection, 445

Line segment, 393

Linear equations, 176
    graphing, 237–242
    of horizontal lines, 239
    of vertical lines, 242
    writing in slope-intercept form, 225–232, 322

Linear functions, 266
    constructing, 273–280
    from graphs, 274–275
    from tables, 276–278
    from verbal descriptions, 279–280
    graphing, 266–268

Linear relationships, 175

Lines
    parallel, 382
    perpendicular, 381

Lines of fit, 591–596, 599–604
    conjectures with, 595–596, 602–604
    drawing, 593–594
    equations for, 599–601

Missing dimensions
    of angles, 386–388, 395–396, 400
    of cones, 561
    of cylinders, 559–560
    of spheres, 562

Mixed numbers
    as decimals, 72

Monomials, 13
    division of, 20
    multiplication of, 16

Natural numbers, 69

Negative exponent, 36–38

Nonlinear functions, 293

Numerical expressions
    evaluating, 7

Order of operations, 7

Outlier, 583

Output, 253

Parallel lines, 382

Perfect cube, 84

Perfect square, 79

Perpendicular lines, 381

Power of a Power Property, 25

Power of a Product Property, 27

Powers, 3
    as products, 3–5
    dividing, 17–18
    multiplying, 14–16
    of a product, 27–28
    product of, 14–16
    quotient of, 17–18
    to a power, 25–26

Preimage, 435

Principal square root, 80

Product of Powers Property, 14

Proof, 413

Proportional relationships, 175–185
    and slope, 176–177
    comparing, 182–186
    direct variation, 213–220
    graphing, 179–182

Pythagorean Theorem, 406–410
    converse of, 417–420
    finding distance with, 423–424
    proof of, 413
    proof of converse, 420

Q

Qualitative graphs, 305–310

Quotient of Powers Property, 17

R

Radical sign, 80

Rate of change, 176, 273

Rational numbers, 69
integers, 69
natural numbers, 69
repeating decimal, 70
terminating decimal, 70
whole numbers, 69

Real numbers, 91–98
classifying, 93–95
comparing, 111–115
ordering, 111, 116–118
rational numbers, 69
irrational numbers, 91

Reflections, 445–452

Relation, 253

Relative frequency, 612–616, 619–624

Remote interior angles, 397–398

Repeating decimal, 70

Right triangle, 405
hypotenuse of, 405
legs of, 405

Rise, 191

Rotations, 455–460

Run, 191

S

Scale factor, 465, 507–508, 517–518

Scatter plots, 581
and lines of fit, 591–596
constructing, 581–582
interpreting, 583–586

Scientific notation, 43–50
addition, 57–58
and technology, 46
computing with, 55–58
division, 56
estimating, 49–50
multiplication, 55
subtraction, 57–58
writing numbers in, 47–48

Similar figures, 205
similar triangles, 205–206

Similar polygons, 513, 517–518

Similarity, 501–504

Angle-Angle (AA) Similarity, 515–516

Similarity statements, 514

Slope, 176–178, 191–200
and equations for lines of fit, 599–601, 603–604
and direct variation, 213–214
and proportional relationships, 176
and systems of equations, 332
formula, 196
of linear functions, 273
undefined, 198, 200
zero, 198–199

Slope-intercept form, 225–232, 237–242
constructing linear functions, 273–280
equations for lines of fit, 599–604
graphing using, 237–239

Slope triangles, 206–210

Solution
number of, 159–164, 322–336
of a linear equation, 176
of a system of equations, 319

Spheres, 551
missing dimensions of, 562
volume of, 551–553

Square roots, 79–82
estimating, 102–103
solving equations, 83

Standard form, 43–45

Substitution method, 341

Systems of equations, 319
number of solutions, 320, 323, 325, 332–334
solving by elimination, 351–358
solving by graphing, 320–326
solving by substitution, 341–346
writing, 363–370

T

Term, 13

Terminating decimal, 70

Transformations, 435
and congruence, 481–490
composition of, 483
dilations, 465–470
identifying for congruent figures, 489–490

identifying for similar figures, 505–506
reflections, 445–452
rotations, 455–460
translations, 435–440

Translations, 435–440

Transversal, 382

Triangle, 393
exterior angles of, 397–400
right triangle, 405
sum of interior angles, 394–395

Truncating, 105–106

Two-way table, 609
associations in, 619–624
constructing, 609–611
relative frequencies in, 612–616, 619–624

U

Unit rate, 175

V

Vertex, 393

Vertical line test, 257–258

Volume
of composite solids, 568–570
of cones, 543–546
of cylinders, 535–538
of hemispheres, 554
of spheres, 551–553

W

Whole numbers, 69

Y

Y-intercept, 225
and equations for lines of fit, 599–601, 603–604
and systems of equations, 332
of linear functions, 273

Z

Zero exponent, 33–35

Zero Exponent Rule, 33

Selected Answers

Lesson 7-1 Angle Relationships and Parallel Lines, Practice Pages 391–392

1. alternate interior **3.** alternate exterior
5. $m\angle 2 = 60°$; Since $\angle 1$ and $\angle 2$ are alternate interior angles, they are equal. $m\angle 3 = 120°$; Since $\angle 2$ and $\angle 3$ are supplementary, the sum of their measures is 180°. **7.** alternate exterior angles; $m\angle 2 = 108°$ **9.** $x = 7$; Corresponding angles are congruent, so $4x = 3x + 7$. Solving the equation for x gives $x = 7$. **11.** Sample answer: Interior angles that are on the same side of the transversal are supplementary. One of the interior angles is supplementary to the corresponding angle of the other interior angle. **13.** 63°; Sample answer: The measure of either of the angles adjacent to the 148° angle is 32° because it is supplementary to 148°. $85° + W + 32° = 180°$, so $m\angle W$ is 63°.

Lesson 7-2 Angle Relationships and Triangles, Practice Pages 403–404

1. $x = 120$ **3.** $m\angle F = 40°$, $m\angle G = 40°$, $m\angle H = 100°$ **5.** $m\angle 1 = 31°$, $m\angle 2 = 106°$
7. $m\angle ADC = 100°$, $m\angle DCB = 39°$ **9.** Sample answer: After finding the value of x, the value should have been substituted into each expression. $4(12) = 48$, $7(12) = 84$. So, the three angles measure 48°, 48°, and 84°.
11. false; Sample answer: If an angle of a triangle is obtuse, then the exterior angle that is supplementary to it will be an acute angle.

Lesson 7-3 The Pythagorean Theorem, Practice Pages 415–416

1. 20.8 in. **3.** 16 in. **5.** 36 cm **7.** 1 more hour **9.** 6.9 cm **11.** Sample answer: When using the Pythagorean Theorem, the square of each leg should have been found, not the sum of the legs and then taken the square. The correct use of the formula is $20^2 + 20^2 = c^2$ and $c \approx 28.3$.

Lesson 7-4 Converse of the Pythagorean Theorem, Practice Pages 421–422

1. no; $45^2 + 56^2 \neq 72^2$ **3.** yes; $12^2 + 5^2 = 13^2$
5. yes; Since $4^2 + 4^2 = (4\sqrt{2})^2$, the angle is a right angle and therefore square. **7.** yes; $\left(7\frac{1}{2}\right)^2 + 18^2 = \left(19\frac{1}{2}\right)^2$ **9.** yes; Sample answer: The remaining distance along the base is 12 feet. The triangle formed by the sides 5 ft, 12 ft and 13 ft form a right triangle. Therefore, the segment he measured is perpendicular to the base and therefore the height. **11.** Sample answer: If you know the lengths of two adjacent sides, and the length of the diagonal, you can use the converse of the Pythagorean Theorem to determine if the angle between the adjacent sides is a right angle. If it is, then the parallelogram is a rectangle. **13.** Sample answer: If he adds 2 inches to each side, the new triangle will have sides 10 inches, 17 inches, and 19 inches. These three values do not satisfy the Pythagorean Theorem because $10^2 + 17^2 \neq 19^2$.

Lesson 7-5 Distance on the Coordinate Plane, Practice Pages 427–428

1. 7.2 units **3.** 5 units **5.** 4.5 units
7. about 63 feet; Rosa is 4.24 units, or 424 feet, from the shelter house and she is about 3.61 units, or 361 feet, from the fossil exhibit.
9. 25 square units **11.** false; Sample answer: If the two points create a horizontal or vertical line segment, the Pythagorean Theorem is not needed.

Module 7 Review Pages 431–432

1. B **3.** Angles *J* and *K* have equal measures. The angles form an isosceles right triangle. The measure of ∠*L* is twice the measure of ∠*K*.
5. B

7.

Ramp Dimensions	Right Triangle?	
	yes	no
3, 4, and 5 feet	X	
5, 10, and 12 feet		X
9, 12, and 15 feet	X	
10, 15, and 20 feet		X

Lesson 8-1 Translations, Practice Pages 443–444

1. *A'*(1, 0), *B'*(0, −3), *C'*(3, −1)

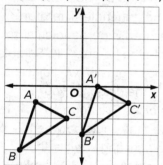

3. (*x, y*) → (*x* + 7, *y* − 4); *Q'*(5, −2), *R'*(4, −8), *S'*(8, −6) **5.** (*x, y*) → (*x* + 2, *y* − 5)
7. The distance from the school to the park is approximately 11.3 units. **9.** Sample answer: The figure is in the same position as the original figure. Since 3 and −3 are opposites, and −4 and 4 are opposites, the translations cancel each other. **11.** Sample answer: For the figure to be translated, every point or vertex must move the same distance and in the same direction. Therefore, the vertices cannot move different distances or different directions.

Lesson 8-2 Reflections, Practice Pages 453–454

1. *A'*(−3, −4), *B'*(1, −4), *C'*(3, −1)

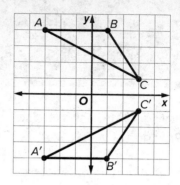

3. *C'*(4, 2), *D'*(8, −2), *E'*(10, 6)

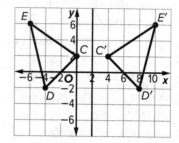

5. (*x, y*) → (−*x, y*); *T'*(0, 3), *U'*(3, 0), *V'*(4, 4)
7. *X'*(−2, −2), *Y'*(−3, 4), *Z'*(1, 2) **9.** Sample answer: He found the coordinates after a reflection across the *x*-axis. The coordinates should be *W'*(−2, 2), *X'*(−2, 4), *Y'*(−4, 4), and *Z'*(−4, 2). **11.** never; Sample answer: When reflecting a figure, the preimage and image are congruent, so, they are the same size and shape.

Lesson 8-3 Rotations, Practice Pages 463–464

1. *E'*(3, 1), *F'*(4, −1), *G'*(3, −3), *H'*(0, 0)

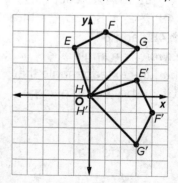

3. (*x, y*) → (−*x, −y*); *Q'*(2, −2), *R'*(3, 4), *S'*(−1, 2)
5. (*x, y*) → (*y, −x*,); 90° **7.** *A''*(7, −4), *B''*(7, −12), *C''*(0, −12), *D''*(0, −4) **9.** Yes; Sample answer: A figure can rotate a total of 360°. A clockwise rotation of 270° leaves 90° remaining in the

rotation before returning to its original position. The original figure could be rotated 90° in the opposite direction, counterclockwise, and be in the same position as the figure rotated 270°. **11.** never; Sample answer: A figure and its rotated image are congruent, so they will always have the same area and perimeter.

Lesson 8-4 Dilations, Practice Pages 473–474

1.

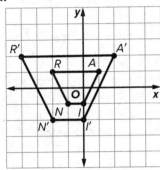

3.

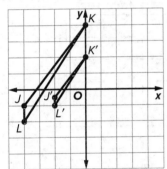

5. $(x, y) \rightarrow \left(\frac{1}{4}x, \frac{1}{4}y\right)$ **7.** $1,610.28 **9.** no; Sample answer: Both coordinates of all points must be multiplied by the same scale factor. **11.** sometimes; Sample answer: A dilation results in an image that is similar to the preimage. It will always be the same shape. A dilation by a scale factor other than 1 results in an image with a different size. The preimage and image will be the same size and shape if the scale factor is 1.

Module 8 Review Pages 477–478

1A.

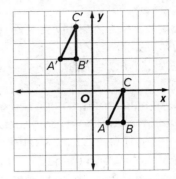

1B. $A'(-2, 2)$, $B'(-1, 2)$, $C'(-1, 4)$ **3.** $\triangle A'B'C'$ is the image of a reflection of $\triangle ABC$ across the y-axis. The x-coordinates of the vertices of $\triangle ABC$ and its image $\triangle A'B'C'$ are opposites. The y-coordinates of the vertices of $\triangle ABC$ and its image $\triangle A'B'C'$ are the same. **5.** C **7.** The coordinates of $\triangle Q'R'S'$ are half that of the coordinates of $\triangle QRS$. The dilation is a reduction. The coordinate notation for the dilation is $(x, y) \rightarrow \left(\frac{1}{2}x, \frac{1}{2}y\right)$.

Lesson 9-1 Congruence and Transformations, Practice Pages 491–492

1. not congruent; Sample answer: No sequence of rotations, reflections, and/or translations will match the two figures up exactly. **3.** Sample answer: If you rotate parallelogram *CAMP* 90° counterclockwise about the origin and then translate it 4 units down, it coincides with parallelogram *SITE*. **5.** Sample answer: a reflection followed by a translation; They are congruent. **7.** yes; Sample answer: A 180° clockwise rotation about the origin maps $\triangle ABC$ onto $\triangle A''B''C''$. **9.** Sample answer: The two trapezoids are not congruent because no sequence of translations, reflections, and/or rotations will map trapezoid *ABCD* onto trapezoid *EFGH*.

Lesson 9-2 Congruence and Corresponding Parts, Practice Pages 499–500

1. $\angle W \cong \angle X$, $\angle U \cong \angle Y$, $\angle S \cong \angle T$, $\overline{WU} \cong \overline{XY}$, $\overline{US} \cong \overline{YT}$, $\overline{SW} \cong \overline{TX}$ **3.** $\angle F \cong \angle H$, $\angle FGE \cong \angle HGJ$, $\angle E \cong \angle J$, $\overline{FG} \cong \overline{HG}$, $\overline{FE} \cong \overline{HJ}$, $\overline{EG} \cong \overline{JG}$ **5.** 62° **7.** 5.3 ft
9. Sample answer: An attic roof is composed of two congruent right triangles. The hypotenuse of one of the triangles is 17 feet. What is length of the hypotenuse of the second triangle?
11. true; See students' work for sample triangles.

Lesson 9-3 Similarity and Transformations, Practice Pages 511–512

1. similar; Sample answer: Dilating rectangle *ABCD* using a scale factor of 0.5 and center of dilation at the origin, and then translating it 3 units to the right maps rectangle *ABCD* onto rectangle *EFGH*. **3.** Sample answer: Dilate triangle *ABC* using a scale factor of 2 and center of dilation at the origin, and then rotate it 90° counterclockwise about the origin. **5.** A
7. no; Sample answer: The area of the final image is 60 square inches. **9.** See students' responses.

Lesson 9-4 Similarity and Corresponding Parts, Practice Pages 521–522

1. not similar **3.** similar; $\triangle YWX \sim \triangle UWV$
5. 14 units **7.** The triangles are similar. $\triangle ABC \sim \triangle XYZ$ **9.** 54.4 mm **11.** See students' work for sample triangles. **13.** 18 in.

Lesson 9-5 Indirect Measurement, Practice Pages 527–528

1. 5.25 ft **3.** 2.25 ft **5.** 20 ft

7.

Person/ Item	Shadow Length (ft)	Height of Person/Item (ft)
Mr. Nolan	9	6
Flagpole	48	32
School	63	42
School Bus	16.5	11

9. true; Sample answer: The triangles are similar using Angle-Angle Similarity.
11. Sample answer: The student set up the proportion incorrectly. One correct proportion is $\frac{h}{5} = \frac{25}{20}$. The correct height is 6.25 feet.

Module 9 Review Pages 531–532

1A. yes **1B.** Sample answer: A translation 4 units right, followed by a reflection across the *x*-axis maps quadrilateral *ABCD* onto quadrilateral *JKLM*.

3.

Statements	Correct	Incorrect
$\overline{AB} \cong \overline{WX}$	X	
$\overline{BC} \cong \overline{WZ}$		X
$\overline{AD} \cong \overline{XY}$		X
$\angle B \cong \angle X$	X	
$\angle A \cong \angle W$	X	
$\angle C \cong \angle Z$		X

5A. no **5B.** Sample answer: $\frac{AB}{JK} = \frac{4}{8}$ or $\frac{1}{2}$; $\frac{AC}{JL} = \frac{3}{5}$; and $\frac{BC}{KL} \approx \frac{5}{9.4}$. Since the ratios between the side lengths are not equal, the two triangles are not similar. **7.** 18

Lesson 10-1 Volume of Cylinders, Practice Pages 541–542

1. 3,078.8 cm³ **3.** 192π in³ **5.** 15.6 ounces
7. B **9.** about 1.4 hours **11.** Sample answer: She used the diameter in the calculation instead of the radius. The volume is $\pi(4^2)(23)$ or 1,156.1 in³. **13.** yes; Sample answer: The cylinder has a volume of about 3,418 cm³. The volume of the prism is 400 cm³. Since 3,418 > 400, the water will overflow.

Lesson 10-2 Volume of Cones, Practice Pages 549–550

1. $130\frac{2}{3}\pi$ ft³   3. 75π mm³   5. 19.3 in³
7. 42.5 in³   9. cost of the yogurt in cylinder: 28.3 · $0.10 = $2.83; cost of the yogurt in cone: 9.4 · $0.10 = $0.94; difference in the cost: $2.83 − $0.94 = $1.89   11. height of 4 inches and radius of 6 inches   13. three times; Sample answer: If a cone and a cylinder have equal base areas and equal heights, the volume of the cylinder is three times that of the cone. So, if the volumes are equal, the height of the cone must be three times that of the cylinder in order for the volumes to be equal.

Lesson 10-3 Volume of Spheres, Practice Pages 557–558

1. $16,222\frac{2}{3}\pi$ ft³   3. 38.8 mm³   5. 629.9 ft³
7. 1.8 mm³   9. 13.1 cm³   11. Sample answer: You could multiply by 4, then divide by 3.
13. Luci; Sample answer: By keeping the volume in terms of π, her answer is closest to the exact volume. Because Stefan used an approximation for π, the volume he found is an approximation.

Lesson 10-4 Find Missing Dimensions, Practice Pages 565–566

1. 2 ft   3. 12 ft   5. 27 in.   7. 5.4 in.   9. 2 more
11. Sample answer: The student did not multiply by the reciprocal of $\frac{4}{3}$ and also did not take the cube root, but instead divided by 3. The radius is 15 in.   13. Cylinder B; It is 3.45 inches taller.

Lesson 10-5 Volume of Composite Solids, Practice Pages 573–574

1. 1,922.7 in³   3. 2,198.9 cm³   5. 14.5 in³
7. 78.5% or 78.6%   9. 71 cones   11. Sample answer: Since both volumes include π, Mateo can find the total volume using $(250 + 25)\pi$. The total volume is 275π ft³.

Module 10 Review Pages 577–578

1. 326.7   3. 144π cubic centimeters   5. 523.6
7. 3

Lesson 11-1 Scatter Plots, Practice Pages 589–590

1. Sample answer:

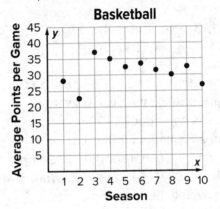

Basketball

3. B   5. Sample answer: The scatter plot would show a positive association. As the temperature increases, the cups of lemonade sold also increases.   7. Sample answer: The scatter plot could represent the relationship between the number of games played and the total points scored. As the number of games played increases, the total number of points scored each game increases.

Lesson 11-2 Draw Lines of Fit, Practice Pages 597–598

1. Sample answer: Since most of the points lie close to the line, the model is a good fit.

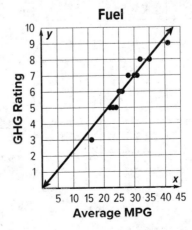

Fuel

3. Sample answer: 27 cups of hot chocolate
5. yes; Sample answer: Since most of the points lie close to the line, the model is a good fit. **7.** sometimes; Sample answer: A line of fit can only be used to make a prediction if there is a positive or negative association to the data. If there is no association between the data then you cannot draw a line of fit and use it to make a prediction.

Lesson 11-3 Equations for Lines of Fit, Practice Pages 607–608

1. Sample answer: $y = 0.0375x + 2.8$; The number of girls participating in high school sports increases by about 37,500 girls every year. In 2000, about 2.8 million girls participated in high school sports. **3.** Sample answer: $y = x + 30$; 70 degrees **5.** about $44 **7.** Sample answer: If the scatter plot has a positive correlation, then the slope will be positive. If it has a negative correlation, the slope will be negative. **9a.** yes; (4, 15); Sample answer: The point of intersection represents the month in which she ate out and ate in the same number of times. **9b.** Sample answer: Replace x and y in the equations for the lines of fit and simplify to see if (4, 15) is a solution of both equations.

Lesson 11-4 Two-Way Tables, Practice Pages 617–618

1.

	Math Club	No Math Club	Total
Chess Club	8	15	23
No Chess Club	11	10	21
Total	19	25	44

3.

	Male	Female	Total
Bus	110; 0.56	84; 0.67	194
No Bus	85; 0.44	42; 0.33	127
Total	195; 1.00	126; 1.00	321

See table for column relative frequencies; Sample answer: Male students are more likely than female students to not ride the bus because 0.44 > 0.33.

5.

	Sit-Down	Suspended	Total
Loops	6	5	11
No Loops	17	3	20
Total	23	8	31

Lesson 11-5 Associations in Two-Way Tables, Practice Pages 627–628

1.

	7th	8th	Total
Attending	80; 0.37	138; 0.63	218; 1.00
Not Attending	105; 0.52	97; 0.48	202; 1.00
Total	185	235	420

Sample answer: The data suggest that there is an association between attendance and grade, because the relative frequencies are different. A 7th grade student chosen at random is more likely to not attend the dance as to attend.

3. A **5.** yes; Sample answer: There is an association between riding a roller coaster but not a water ride because the relative frequencies are different. A student chosen at random that rode a roller coaster is less likely to have also ridden a water ride.

7. Sample answer: There are 30 students in Jenna's homeroom. Of the 14 girls, 6 ride the bus to school. There are 5 boys that walk to school. The data suggest an association because the relative frequencies are different.   **9.** Sample answer: Find the relative frequencies by row or column. Then write a proportion to make a prediction.

Module 11 Review Pages 631–632

1A. Sample answer:

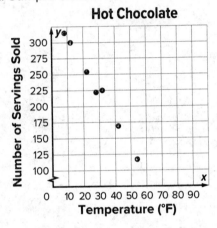

Hot Chocolate

1B. B   **3.** Sample answer: about 225 riders

5.

	Pet	No Pet	Total
1st	18	12	30
2nd	18	9	27
3rd	14	9	23
Total	50	30	80